Innovations in the European Economy between the Wars

Innovations in the European Economy between the Wars

edited by

François Caron, Paul Erker, Wolfram Fischer

W DE G

Walter de Gruyter · Berlin · New York

1995

"This publication is the result of international research carried out between 1989 and 1993 within a Scientific Network on the *Economic History of Europe between the Wars* supported by the European Science Foundation".

Printed on acid-free paper which falls within the guidelines of the ANSI to ensure permanence and durability.

Library of Congress Cataloging in Publication Data

Innovations in the European economy between the wars / edited by François Caron, Paul Erker, Wolfram Fischer.
p. cm.
"The result of international research carried out between 1989 and 1993 within a Scientific Network on the 'Economic history of Europe between the wars' supported by the European Science Foundation" – T.p. verso.
Includes bibliographical references and index.
ISBN 3-11-013582-5
1. Technological innovations – Economic aspects – Europe-History – 20th century. 2. Europe – Economic conditions – 1918–1945. 3. Research, Industrial – Europe – History – 20th century. 4. Technological innovations – Economic aspects – Europe – Case studies. 5. Research, Industrial – Europe – Case studies. 6. Diffusion of innovations – Europe – Case studies. 7. Industrial organization – Europe – Case studies. 8. Comparative organization. I. Caron, François. II. Erker, Paul. III. Fischer, Wolfram.
HC240.9.T4I546 1995
338'.064'094–dc20 95–1423
CIP

Die Deutsche Bibliothek – Cataloging in Publication Data

Innovations in the European economy between the wars / François Caron ... (ed.). – Berlin ; New York : de Gruyter, 1995
ISBN 3-11-013582-5
NE: Caron, François [Hrsg.]

Printed in Germany
Text conversion and printing: Arthur Collignon GmbH, Berlin
Binding: Lüderitz & Bauer-GmbH, Berlin

Contents

III. The Structure of Enterprise and Innovation

IV. The Formation and Development of Industrial Research and Technology

List of Contributors

Hans-Joachim BRAUN, University of Armed Forces, Hamburg, Department of Social and Economic History, and the History of Technology
John CANTWELL, University of Reading, Department of Economics
François CARON, University of Paris, Institute of History
Bernd DORNSEIFER, Free University of Berlin, Department of History
David EDGERTON, University of Manchester, Center for the History of Science, Technology and Medicine
Paul ERKER, Free University of Berlin, Institute for Social and Economic History
Wolfram FISCHER, Free University of Berlin, Institute for Social and Economic History
Renato GIANNETTI, University of Florence, Department of History
Pascal GRISET, Center for National Research Studies, Paris, Institute for Modern and Contemporary History
Sally M. HORROCKS, University of Lancaster, Department of History
David A. HOUNSHELL, Carnegie Mellon University, Pittsburg, Department of History
Gottfried PLUMPE, Bayer AG, Leverkusen, Public Relation
Nathan ROSENBERG, Stanford University, Department of Economics
Harm SCHRÖTER, Free University of Berlin, Institute for Social and Economic History
Francis SEJERSTED, University of Oslo, Center for Technology and Culture
Peter TEMIN, Massachusetts Institute of Technology, Department of Economics
Nick VON TUNZELMANN, University of Sussex, Science Policy Research Unit

List of Tables

List of Figures

Preface

WOLFRAM FISCHER

At the invitation of the European Science Foundation in April 1989 twenty economic historians from fourteen European countries, Canada and the United States of America gathered in the premises of the Foundation at Strasbourg to discuss the creation of a network on the "Economic History of Europe Between the Wars". The initiative came from professor François Caron, Paris. The assembled scholars identified three topics which promised interesting new results:

1. International monetary and financial developments,
2. The source and diffusion of technical change,
3. The working of the labour market.

A Coordination Committee of nine scholars from seven European countries and the United States was elected to be chaired by François Caron; it met several times in France to work out the schemes for three work shops in different European countries and a final conference in Paris. The workshop on the first topic, organized by Charles Feinstein, Oxford, took place in May 1992 at Venice; the workshop on the third topic, organized by Rolf Ohlsson, Lund, took place in May 1992 at Lund.

The workshop on the "Sources and Diffusion of Innovation", as it finally was called, organized by Wolfram Fischer, Berlin, took place at the *Historische Kommission zu Berlin* in July 1991. Seventeen economic historians and economists from six European countries and the United States took part. Unfortunately, experts from Eastern European countries could not be recruited; the South was only respresented by one Italian, the North by one Norwegian scholar. Most participants came from the three "big" European countries France, Germany and Great Britain – and from the USA. This seems to reflect, however, the state of the arts in this field which until recently was dominated by scholars from the United States.

Two of them were asked to comment on the papers, another one, David Hounshell, to give a paper.

This book presents the results of the Berlin workshop. The papers and comments given in 1991 were re-written in the light of the discussion and those of none-English speakers were edited by Sally Horrocks, an English member of the group. An introduction by François Caron, the motor of the network and particularly of the topic on technical change, and a summary by Paul Erker, who served as one of the editors of this book, were added.

It may be mentioned that this book can also be regarded as a historical companion to the ambitious work by Horst Albach, Culture and Technical Innovation. A Cross-Cultural Analysis and Policy Recommendations (The Academy of Sciences and Technology in Berlin, Research Report 9) which was published by Walter de Gruyter early in 1994. That work is the result of an interdisciplinary research group, set up by the *Akademie der Wissenschaften zu Berlin* in 1987 which was chaired by Wolfram Fischer while its motor, *spiritus rector*, and final author was Horst Albach. At least the Berlin members of the workshop of the European Science Foundation have learnt a lot from the economists, engineers, chemists, lawyers, sociologists, psychologists, industrialists, and Japan-experts who debated that topic over several years with particular reference to the contemporary German, American, and Japanese economies. The present book adds a Western European and historcial dimension to this dicussion.

The editors thank the European Science Foundation for funding the network and the conference; they thank particularly Dr. John H. Smith, Dr.Gérald Darmon, and Ms. Margaret Kinane for their administrative assistance, and all the participants for their continuous work on the topic even several years after the workshop had taken place. They are also grateful to the publisher who has agreed to publish this book without subsidy in a time when conference-volumes are exceedingly difficult to sell.

Introduction[1]

FRANÇOIS CARON

Any history of the economic growth between the wars must take into account the achievements of technology. This period marks a decisive step in the sectoral deepening and spatial widening of the process of industrialisation. The new techniques which appeared during the last quarter of the nineteenth century, and which were to form the basis of the consumer society, achieved ever wider fields of application, not only in those countries which became industrialised early, but also in those which had but recently embarked on the path of industrial modernity. It would, however, be absurd to treat the problems of development in the 1920s and '30s in the same terms as those of the 1820s and '30s, since the techniques which served to propagate it had changed and did not have the same effects on either economic growth or economic structures.

In reality, a new model of growth was in the making, which was preparing for the achievements to come after the Second World War. The strong economic growth which took place in Europe in the 1950s and '60s was not simply the result of a process of "catching up", but was also part of a continuous effort of deepening and rationalisation, begun in the 1920s and pursued during the 1930s, in ways and conditions which varied considerably from one decade to the next. However, the transitory nature of the experience makes it all the more difficult to analyse. We shall attempt to arrive at an understanding of the complexity of these developments using three different approaches:

i) by defining the original characteristics of economic growth in the interwar period;
ii) by clarifying the nature of the structural changes which occurred, including unemployment in the analysis;

[1] Translated from French by Elizabeth Aitam.

iii) by analysing the dynamics of the technical system, which also explain the other factors, taking into account not only the role of "interdependence", but also by investigating the strategy of firms in the area of innovation.

1. *Defining Growth*

Placed in a long-term perspective, the performance of European economies in the interwar period appears on the whole favourable. The real problem lies in understanding why growth won over stagnation, despite the accumulation of numerous and serious obstacles. This observation results as much from an analysis of the evolution of production as of productivity.

1.1. Rates of Production

If one examines the years 1924–37, as Charles Feinstein has done for the United Kingdom, it would appear that many countries attained rates of growth in national product of more than 2% per annum, some even reaching growth of over 3%: amongst those achieving more than 3% we find countries such as Sweden, Norway, Germany and Italy, but also Hungary and Bulgaria. The United Kingdom, Belgium, Denmark, Czechoslovakia, Spain (from 1924–35) and Switzerland had growth rates of between 2 and 3%, while Austria, France (between 1924–38) and Holland did not even reach 2%, or even 1% in the case of the first two countries mentioned. These three countries suffered much from the prolongation of the depression of the 1930s; even in 1937–38 their domestic national product remained considerably smaller than it had been in 1929. These facts illustrate the harmful effect of the deflationary monetary policies of the early 1930s.

These data also invite two further steps of analysis: we need to compare them with the figures for the preceding and following years, and at the same time, to compare the performance of the 1920s with that of the 1930s. The British experience is particularly interesting. In the years from 1896 to 1913 this country did not experience the acceleration in growth which took place in many other European countries, and indeed it went through a serious recession between 1913 and 1924. However, between 1924 and 1937 the growth in gross domestic product (GDP) was clearly higher than it had been in the years 1873–1913 (2.2% compared to 1.8%).

It is as though the entire interwar period was a preparation for the rapid growth to come after the war (2.8% from 1951 to 1973). In countries such as France, Spain, Italy, Hungary and Scandinavia, which had experienced rapid growth in the ten or twenty years preceding the 1914 war, this tendency persisted and in some cases increased in the 1920s. In the 1930s growth was either interrupted or at least greatly reduced, as in France, or continued, as in Scandinavia and, albeit according to an entirely different system, in Italy and Germany.

In the vast majority of cases, growth in the industrial sector exceeded that in GDP and with more noticeable fluctuations; in the United Kingdom, for an average growth in GDP of 2.3%, industrial growth reached 3.2%. Within the industrial sector the differences are marked. For countries which were members of OECD in 1955, growth in industrial production (manufacturing) was 3.1%, but, on the basis of aggregate data, it ranges from a minimum of 1% in the textile industry to a maximum of 4.3% in the chemical industry. Growth was 3.9% and 3.6% in the metal products and basic metals industries respectively. In the food, beverages and tobacco industries it reached only 1.9%. Growth was smallest therefore in the sectors of consumption of non-durables and semi-durables. Durable goods were certainly gaining a position of ever greater importance in the production system. In the same way, the European car industry was undergoing remarkable expansion, and from 1926 to 1937 production in the OECD member countries grew at a rate of 7.7% per annum. It was nevertheless the semi-finished products (chemistry and basic metals) and capital goods industries which, on the whole, experienced the largest growth.

This observation suggests that technical change in the industrial sector constitutes the main explanatory factor for the dynamism of growth in an economy subject to unstable and uncertain demand. The opportunities offered by industrial technology compensated for the depressing effects of expectations founded on financial data and on demand. In addition, in the second half of the 1930s, other, external factors intervened which were linked both to the preparations for war and to voluntarist industrialisation policies.

1.2. Rates of Productivity

This hypothesis appears to be confirmed by the analysis of performance in terms of productivity. The interwar years fall within a long period of growth in labour productivity. Calculations based on production by man-year rather than by man-hour tend, particularly in this period, to obscure the importance of this growth. The evolution in France is quite characteristic from this point of view: growth in labour productivity, calculated in man-hours, had reached 2.0% per annum between 1896 to 1913, whereas it reached 2.4% between 1924 and 1938. A comparison between the 1920s and the 1930s is also most instructive. In many countries, in fact, growth in productivity in the 1930s was slightly faster or equal to what it had been in the 1920s. Indeed, when it was slower, the gap between the two figures was always much greater for production than for productivity. In France, growth in production in all sectors fell from 2.8% to –0.5% from 1924–29 to 1929–38. Despite this fall, productivity continued to grow in the 1930s at a rate only slightly slower than that of the 1920s: 2.1% instead of 2.9% in all sectors taken together, 2.9% instead of 3.4% in industry. In France, as in most European countries, the laws introducing a reduction in working hours were passed immediately after the war and wage costs increased considerably. In addition, in the second half of the 1920s French industry lacked manpower. The result of this was an attempt at rationalisation which persisted into the 1930s, albeit in a rather different form, as we shall see, in the sense that it was above all necessary, this time, to combat the shrinking of the market and of profits caused by the Depression.

These developments were not unique to France. Charles Feinstein's conclusions about Total Factor Productivity (TFP) in the United Kingdom offer us a better perception of the specificity of the period under examination. Between 1924 and 1937 the TFP of GDP grew at a rate of 0.7% per annum as against 0.45% between 1873 and 1913. However, real industrial performance is much more impressive: here the rate passes from 0.6% to 1.9%, i. e. it triples. A comparison between the 1920s and 1930s is also informative: for a growth in GDP of 2.6% between 1924 and 1929, the TFP grew at a rate of 1.2%, whereas from 1929 to 1937 the rates were at 2% and 0.6% respectively. In addition, the TFP of the industrial sector increased more rapidly in the years 1929–37 than in the years 1924–29 (2.4% and 1.8%). The sector which realised the best performance was the textile industry with a rate of growth in TFP of 4.4%: this figure illustrates in a remarkable way the effect of the rationalisation policies imposed by the Depression.

One must conclude, with Charles Feinstein, that the factors which explain the growth in TFP are to be found on the supply side. The most important of these was, evidently, the attempt at rationalisation, which was an essential component of technical change. In the case of the United

Kingdom, there was both a catching-up effect in the area of new technology, where the UK had fallen behind Germany and the United States before, and above all during the war, and a process of autonomous technological development of domestic origin[2].

Although this analysis applies to specifically British realities, it also gives an insight into the developments in all European countries. Catching-up and deepening were both brought about by the technologies of the "second industrial revolution". As we have said, it was the perfomance in the two sectors of semi-finished and capital goods which explains the high rates of growth in production and productivity. However, they were in their turn sustained by investments from business and the public utilities. The European economy, swept along by electrification, was engaged in a dual process of renewal and development of its basic investments, and above all of equipment. The extent of the trend is demonstrated as much by data concerning rates of investment, formation and structure of capital as by those related to industrial mechanisation[3]. This trend has an impressive characteristic of a general nature, and it concerns as much the countries of Eastern Europe as Spain, Italy, the Scandinavian countries, France, Germany or the United Kingdom. The collapse which occurred in France in the 1930s is far from being a general phenomenon. In fact, in most countries activity took off again after the Depression in the second half of the 1930s.

In France, Spain and Italy, the growth in rates of productive investment considerably increased the capital stock available per worker. But, generally, the speed of renewal of equipment greatly reduced this growth, in comparison with growth in gross rates of investment. In addition, investments, whether in renewal or in development, brought with them the "capital saving" innovation. Charles Feinstein has shown that there was a tendency towards a decrease in the capital/output ratio between the wars in the industrial sector. He writes:

"It is remarkable that a fall in the capital-output ratio between 1924 and 1937 is found in every manufacturing industry group without exception [...] It is possible to point to technical developments in this period that had, to a greater extent than most innovations have, the effect of reducing capital costs in manufacturing. The chief of these was the

---

[2] See R. C. Matthews/C. H. Feinstein/J. C. Odling-Smee, British Economic Growth, 1856–1973, Oxford 1982, p. 537.

[3] See on this J. J. Carré/P. Dubois/E. Malinvaud, La croissance française, Paris 1972, p. 202, table 12.

changeover to electricity as a source of power, a development that was accompanied by an increase in the proportion of electricity purchased as opposed to generated within the firm. This had its counterpart in the high rate of growth of output and capital stock in the public utilities sector"[4].

Feinstein's analysis can, in our view, be enlarged upon. The technological bias towards capital saving modes of production has a more general significance than that of a simple transfer to the public utilities of the weight of productive investment. For it relates to the whole of the economy. In addition, whereas in the United Kingdom the upward trend in capital productivity did not continue after the World War II, this was not the case, at least until the end of the 1960s, in most industrialised countries. The British economy thus appears to have anticipated, in the interwar period, a model of growth which became typical in Europe in the 1950s and 1960s, marked by the spread of capital saving technologies[5].

2. *Structural Changes and Unemployment*

The growth process was accompanied by structural changes within the working population and an analysis of these changes is the prerequisite to an understanding of unemployment. We shall use the following four themes, each closely linked to the other, as our guide: agricultural under-employment, industrial structures, hidden unemployment in the non-agricultural sectors, and the mechanisms of industrial unemployment.

2.1. Agricultural Under-employment

In most continental countries at the end of the war there existed a large reserve of agricultural manpower, due to a chronic under-employment of labour capacity. In Eastern Europe, as well as in the Mediterranean, the percentage of agricultural workers remained higher than or close to 50% of the total working population and its decline was slow: even without quoting the extreme examples of Yugoslavia and Romania, these levels range from 56% in 1910 to 54% in 1930 in Hungary, and from 59% in 1921 to 52% in 1936 in Italy. Even the Nordic countries (71%, 39%, 36%

---

4 Matthews/Feinstein/Odling-Smee, p. 384 f.

5 See F. Caron, Le résistible déclin des sociétés industrielles, Paris 1985, p. 259.

and 30% respectively Finland, Sweden, Norway and Denmark in 1930) and the northern countries of Western Europe (36% in France in 1931, 29% in Germany in 1933) retained high levels of working population. Switzerland and Belgium, with levels of 21% and 17% respectively in 1930, mark the exceptions. One can therefore acknowledge that in the greater part of the European countryside there existed an "immense reserve of manpower", to use Albert Carreras' expression referring to the Spanish countryside, or, at the very least, that the possibilities of transfer from agriculture to industry were far from being exhausted by the end of the First World War. However, it is clear that this movement was not as large as the growth in industrial production might lead one to suppose. Agriculture retained an abundant workforce.

Albert Carreras has shown that Lewis' model, based on an "unlimited supply of labour", applied quite well to Spain. He observes that between 1920 and 1950, contrary to what had taken place in the previous decade, there existed an inverse relation between the productivity of industrial labour and the size of the active male agricultural population[6]. The former increased between 1920 and 1930 and fell from 1930 to 1950, the latter undergoing the reverse process.

However, Gianni Toniolo maintains that, in Italy's case, Lewis' model needs to be much more precisely defined. For one thing, according to data from several official surveys, agricultural under-employment remained significant, since it can be estimated at a third of the working agricultural population. It also had a chronic feature. It is closely linked to the model of the extended family in an agricultural environment, which permits a redistribution of income amongst its members. The labour market does not therefore function according to the simple logic of a comparison between agricultural and industrial incomes. However, the Italian case is not unique: the agricultural sector throughout Europe is characterised by hidden unemployment on a large scale, which was particularly high in Eastern Europe.

Thus the migration from agricultural to industrial activity does not depend, precisely because of chronic unemployment, solely upon the difference between agricultural and industrial wages. It fluctuates mainly according to the supply of industrial employment. At the beginning of the 1930s this migration was greatly reduced in most European countries, despite the maintenance of high industrial wages and a large fall in agricultural

---

[6] See J. Nadal/A. Carreras/C. Sudria, La economia española en el siglo XX, Barcelona 1991, p. 294.

incomes. Denmark, where industrial production increased at this time, is the exception. In France the movement was more intense in the 1920s, in Germany it was greater in the 1930s.

However, the migration from agriculture to industry is not, as Toniolo says, a "one-stage process". It is a reversible phenomenon which may be compared to shared forms of labour, i. e. extremely mobile and part-time. The transition from agricultural labour to industrial waged employment is the result of a long process. Toniolo shows that in the interwar period the hard core of urban workers who had broken all ties with the countryside was relatively small. Toniolo's analysis has a significance that goes beyond the Italian scene, and it invites one to examine the real nature of industrial employment and the changes it underwent[7].

2.2. Industrial Structures

Industrial growth drew a share of the available agricultural workforce to industry, but the increase in industrial working population as a percentage of the total working population was smaller in most countries, in some cases significantly so, than the share of industrial production as a proportion of GDP. This was due to the rapid increase in industrial productivity and in spite of the reduction in working hours and the winding down of former areas of activity. In Sweden, from 1910 to 1930, the percentage of industrial product rose 11 points, and that of the industrial working population 4 points. In the UK, from 1924 to 1937, the percentage of the industrial working population as a share of total working population remained constant, at a level of 32.9%, whereas the share of the industrial sector as a percentage of total production rose from 30.9% to 34.8% in real terms.

But the stability of feeble growth in the size of industrial population in relation to total population conceals the large-scale structural changes which were taking place both within the working population and in industrial products. The former is characterised by a dual mobility: one sectoral, reflecting the changes in product structure, the other professional, which is explained by the changes in modes of production. The figures for France illustrate this very well: from 1906 to 1931 the number of

[7] See G. Toniolo/F. Piva, Unemployment in the 1930s: The Case of Italy, in: B. Eichengreen/T. J. Hatton (eds.), Interwar Unemployment in International Perspective, London 1988, p. 221 ff.

industrial wage-earners rose by almost 1.8 million[8]. There were 5.4 million of them in 1931 and 68% of these new wage-earners were attracted towards companies employing more than 100 workers, mainly in the engineering, electrical engineering and chemical industries. The growth in industrial productivity thus results from the diffusion of production techniques, which required a certain degree of concentration, and a profound change in the nature of industrial working methods. The promoters of electrification at the end of the 19th century had hoped and predicted that the new technology would encourage a dispersal of labour. The experience of the interwar years does not bear out that expectation.

The upsurge in large factories was not peculiar to France, although one must distinguish between countries such as France and Italy, in which medium-sized factories continued to play a major role, and others like the United Kingdom and Germany, where large establishments dominated. These different structures were set up in the 19th century. The increase in manual labour within large factories is, however, not unique to North-Western Europe, and it is possible to consider the industrialisation of Eastern European countries as "dualist" in the sense that it contrasts labour-intensive establishments, typical of expanding industrial sectors, with very small artisan establishments, characteristic of the traditional sectors. It is in fact the almost total absence of medium-sized factories which sets these countries apart, with the sole exception of Czechoslovakia, which had a richer industrial past.

The development of salaried employment in industry does not necessarily imply permanent employment, in fact one might be tempted to say quite the reverse. The great Parisian car factories and the electrolysis plants in the French Alps experienced considerable turnover in their workforces. These observations concur with Gianni Toniolo's conclusions about Italy. He demonstrates that even amongst the hard core of urban workers, the percentage who had a permanent job (or wanted one!) is small: at Alfa Romeo 56% of the workforce stayed for less than a year, and at Montecatini the figure was 83%.[9]

2.3. Hidden Unemployment in the Services and in Industry

A significant part of the workforce freed from agriculture came to feed the service industries, either because of the development of certain sectors

[8] The return of Alsace-Lorraine to France accounts for 15% of this increase.

[9] See Toniolo/Piva, p. 225.

such as transport or administration, or because of the fairly widespread phenomenon of hidden unemployment in these sectors, most particularly in trade. The excess working population in the commercial sector in France was indeed one of the main themes of analysis to be found in official reports on the national economy after the Second World War. Charles Feinstein has clearly shown this to be the case in the United Kingdom. In both countries it was a response to insufficient demand and industrial unemployment. This was however not only true of France and the UK, but contributed significantly to the reduction in performance in terms of productivity of the European economy as a whole.

The mechanisms of the labour market in certain industrial sectors, as described by Gianni Toniolo in the case of Italy, reveal also the presence of hidden forms of unemployment in industry itself. In France, despite the changes we have described taking place in the 1920s, in the 1930s almost a third of the industrial workforce was still made up either of isolated workers or of "petits patrons" ("little bosses"). Their survival and even revival was favoured by the Depression.

2.4. The Significance of Unemployment

The unemployment statistics should be read in the light of these observations. Gianni Toniolo, using the population censuses, has corrected upwards the estimates of industrial unemployment rates in 1932 and 1935. He has increased them from 15.5% to 40.8% and from 11.5% to 23.2%. But he notes that these figures simply mean that workers' periods of unemployment were almost twice as long in 1932 as in 1935. Generally, according to T. J. Hatton, the duration of unemployment was shortest when turnover was highest[10]. On the other hand, when turnover was low, unemployment tended to be of longer duration. A. Newell and J. S. V. Symons have shown clearly that industrial unemployment was aggravated in the 1930s by a lack of wage flexibility[11]. This cannot be explained either by the role of the unions or by government intervention. It is rather the consequence, according to Hatton, of a new strategy on the part of firms in the management of their staff. He writes: "The interwar labour market

---

10 See T. J. Hatton/B. Eichengreen, Interwar Unemployment in International Perspective: An Overview, in: Eichengreen/Hatton (eds.), p. 35 f.

11 See A. Newell/J. S. V. Symons, The Macroeconomics of the Interwar Years: International Comparisons, in: Eichengreen/Hatton (eds.), p. 61 ff.

in Britain and the US [is] a stage of transition between a high turnover, low employment attachment regime typical of the late nineteenth century and one of lower turnover and greater job attachment of the post-war period"[12].

In the case of France and Germany this hypothesis has yet to be proved. In fact, paternalist practices, the avowed aim of which was the preservation of workers' jobs, played a significant role in these two countries in the industrial system inherited from the 19th century. The transition to a managerial system of management may have had the opposite effect on modes of labour organisation to that defined by Hatton.

One further observation should be made here. A sectoral and regional analysis enables one to establish a relationship, tenuous though obvious, between the levels of unemployment and an industrial taxonomy which takes account of the nature of the technologies concerned. The regional analysis made by Charles Feinstein is illuminating from this point of view. Before 1914 the highest levels of unemployment were recorded in the London area, because London acted as a magnet to workers, yet could not absorb the influx of population. Lowest rates are to be found in Scotland, Wales and in the North of England. In the interwar period, the situation was reversed: the lowest unemployment rates were achieved in London and the South-East, and the highest in Wales, Scotland and the North-East. Unemployment in the UK was thus one of the aspects of the sectoral recomposition of the economy. This judgement applies, of course, to countries other than Great Britain.

3. The Dynamics of the Technical System

We are aware of the uncertainties which weigh upon a quantitative evaluation of innovative activity: the most uncertain are those which use lists such as "major inventions, innovations and discoveries". The most often quoted of these lists is that drawn up by C. Streit in 1949, and it was the use of this source which enabled John Dunning to write that "during the interwar years, the pace of technological advance slowed down, and what progress there was strongly favoured the US economy"[13]. The five-yearly

---

[12] Hatton, p. 36 f.

[13] J. H. Dunning, Changes in the level and structure of international production: the last hundred years, in: M. Casson (ed.), The Growth of International Business, London 1983, p. 109.

average of the number of inventions and innovations, according to Streit, was 38 from 1876 to 1914, and 30 from 1915 to 1939, whereas the percentage for the United States in the whole period went from 40 to 60%.

The statistics for patents deposited in the United States seem to us a more pertinent indicator. These lead one to conclude that the growth movement before the war continued into the 1920s. On the other hand, there was a significant fall in the 1930s[14]. But the number of patents of European origin awarded in the United States grew much more rapidly than the total number of patents awarded in the interwar period. In addition, the rate of increase doubled from 1920–24 to 1933–39 in relation to the previous period (1890–96 to 1920–24): 4% compared to 2.04%, and it was considerably higher in the 1920s than in the 1930s (5% compared to 3.2%)[15]. Dunning explains this increase "by the growing interest of foreign firms in the US market". Such a statement requires substantiation, since interest in the American market does not date from the interwar period.

Dunning considers that "in spite of notable inventions of the interwar period – television, radar, the jet engine, colour photography, several man-made fibres and some antibiotic drugs, for example – these were mainly years of development, adaptation and dissemination of the technological and organizational breakthroughs of the late nineteenth and early twentieth centuries". Dunning's view, which contrasts a prewar period, which discovers, with an interwar period, which develops, does apply to certain areas of technology, but not to all. In fact the new branches of technology which appeared in the 1890s and 1900s, such as electricity and organic chemistry, were indeed developed and diffused during the interwar period. By the eve of the 1914 war, these innovations had achieved a level of maturity which no longer left any doubt about their future, whereas more recent innovations, such as aviation and radio, were still in the early stages of their development. The war revealed the immense potential, which only certain, albeit the major, innovations were able to realise.

John Cantwell analyses the sectoral specialisation of patents awarded to Europeans in the following terms: "In Europe almost all the chemical fields displayed rapid growth in the interwar period, including inorganic

[14] J. Cantwell gives the following figures: the average annual rate of increase in number of patents awarded went from 1.95% between 1890–96 and 1920–24, to 1.74% between 1920–24 and 1927–29, and -0.39% between 1927–29 and 1933–39. See the contribution of J. Cantwell in this book.

[15] Ibid.

chemicals and agricultural chemicals; and in the electrical area a fast rate of development extended to telecommunications, illumination devices and general electrical equipment [...] Related to the European strength in chemicals, there was also a fast technological development in Europe in chemical machinery and equipment, and in the materials technologies used to create non-metallic mineral products (in 1920s), and in rubber and plastic products (in 1930s)." Using the same source, he has sought to determine which were the "comparative technological advantages" of Europe, the sectors in which the share of European patents in the total number of patents in that sector was higher than the same share in the total number of patents. Europe's advantage was clearest in the chemical sector, and within this sector "the greatest European strengths centred in agricultural chemicals, bleaching and dyeing and organic compounds". In addition "the Europeans also had an advantage in the development of electrical equipment [...] Perhaps allied to this, Europe also performed well in the field of professional and scientific instruments, especially in photographic equipment". Finally, "in motor vehicles their comparative advantage was in the field of internal combustion engines and not in vehicles as such".[16]

Cantwell's conclusions concerning the orientations of the European technical system are confirmed by the evolution in the sectoral distribution of industrial products in the different countries of Europe: the rise of the electrical engineering, metallurgical and car industries, of the chemical and materials industries, and the stagnation or relative decline of the food, textile, clothing, wood and paper, and naval armaments industries.

We shall use three approaches, each complementary to the other, to describe the dynamics of the structural changes associated with the transformation of the technical system:[17]

i) The first will be based on two concepts: that of the interdependence between areas of application and technology, and that of firms' product strategy.
ii) Secondly, we shall describe the constraints of rationalisation which were imposed on European companies.
iii) Thirdly, using as yet unsubstantiated data, we shall touch on companies' research and innovation strategies.

---

[16] Cantwell, p. 289 f.

[17] See F. Caron, Histoire économique et dynamique des structures, in: Année Sociologique, 41, 1992, pp. 107–28.

3.1. Interdependence and Product Strategy

The interplay of mutual dependence between the different fields of technology should be considered the main engine of technical change. It both explains the rise of the new branches and describes its downstream effects. That is to say it enables us to understand their progress within the system in place. It seems to us that this particular, and partly autonomous, dynamic of technology played a significant role in the interwar period.

In order to understand the dynamic of interdependence from a global point of view, we shall use three concepts, which we shall support with concrete examples: these concepts are demand for invention, "spin-off effect", and technology trajectory. To illustrate the first we shall recall the history of the development of interconnected electrical networks in France, to illustrate the second that of the relationship between the aeronautical and aluminium industries, and lastly radio to illustrate the third concept.

Since before the 1914 war French electrical engineers had, in the main, considered the interconnection of electrical networks to be both possible and desirable. With the advent of war it became an issue of national importance[18]. For the implementation of such a programme the engineers were possessed of a coherent scientific doctrine and expertise which they were able, thanks to an experimental approach based on the idea of "test networks", to adopt to any particular situation which might actually arise[19]. A large measure of consensus had been achieved in this area. Confident of such certainties, the engineers were able to seize the opportunity to carry out any element of the programme as soon as it presented itself. For the very diversity of the technical methods adopted by the existing networks made an immediate global implementation impossible. One should note, however, that since 1918 agreement had been reached on the choice of 50 cycles, whereas arguments still raged about the choice of voltages.

In fact it was pressure from the consumers of electricity which determined which networks were set up first. Two consumer groups played a particularly important role: the producers of aluminium (and more generally electrochemists and electrometallurgists), and the railway companies. The electrolysis technology used in French aluminium factories, which

---

[18] See Histoire de l'électricité en France. Tome premier 1881–1918, Paris 1991.

[19] See G. Ramunni, L'élaboration du réseau électrique français. Un débat technique de l'entre deux guerres, in: Un Siècle d'Electricité en France 1880–1980, Paris 1987, pp. 269–91.

was operated in the French Alps by the two great producers of this metal, AFC (or "Compagnie de Produits Chimiques et Electrochimiques, Alais, Froges et Camargue") and the Société d'Electro-Chimie, did not develop in any radical sense between the end of the 1890s and 1920. The functioning of the tanks was less than satisfactory: the intensity of current did not exceed 10,000 to 20,000 amperes, the energy produced was feeble, and working conditions were inhuman. A major technological breakthrough took place at the end of the 1920s with the adoption of two new procedures, one borrowed from electrothermal technology, the Söderberg process of Swedish origin, and the other, know as "brasquage en blocs serrés", which was perfected in French factories after a long process of testing. In the 1930s these two methods were developed concurrently in an attempt at rationalisation, which made possible the closure of several old factories and the concentration of production in new plants. Tanks using a current of 50,000 amperes were put into production in 1934 and achieved a much more satisfactory output. Thus began the process which was to culminate in the 280,000 amperes achieved in 1986 at Saint-Jean de Maurienne.

The changes which had taken place in the methods of aluminium production brought about a radical modification in the methods of electricity production. Until the 1920s electricity supply was maintained by "on-stream" factories, whose production was intermittent and irregular. In the low season the factories had to be closed, which led to an under-utilisation and rapid deterioration of the equipment. The introduction of the new tanks necessitated a vast programme of development in the Alps, which was designed to assure continuity of production. It involved the construction of a series of large dams, and the creation of interconnected electricity plants with links to the factories, which was made possible by the setting up of a high tension network and converter groups.

The only previous significant programme of electrification had been that of the railway network in the Midi before the 1914 war, in the construction of the lines across the Pyrenees. Its engineers had been inspired by the Swiss experiment and had adopted the single-phase current at 60,000 volts. 327 kilometres were electrified in 1913. In 1918 a commission was set up, which reported in 1920, and predicted the construction of 9,000 km over the three networks with hydroelectric power, the Paris-Lyon-Mediterranean, the Paris-Orléans and the Midi networks. The commission selected the direct current option at a high tension of 1,500 volts, this time using the British model, which obliged the Midi network to change the equipment of its lines.

The case of France is not unique: all the European electrical networks were conceived and built as much in response to the requirements of their large consumers as for reasons of rationalising the operation of the electrical network itself. The two motives are inseparable and this linkage illustrates the logic of interdependence. However, the construction of interconnected networks encountered three major obstacles: the diversity of the technical methods previously adopted, which added considerably to investment costs; the as yet insufficient demand to justify operations of this size; and lastly the impediment of ill-adapted legal systems. Writing about Italy, Renato Giannetti says: "It was the relative lack of demand which rendered the construction of large plants rather unappealing, given that to be economical they would have had to feed many centres of consumption", and "the expenditure required for the replacement of machinery, both in power stations and users' homes which would be necessary in the event of unification of frequencies, was considered unsustainable[20].

Hans-Joachim Braun and David Edgerton insist strongly on the importance of the links between the techniques of the aeronautics industry and other sectors of activity, and particularly on the role played by aluminium alloys in this sector[21]. We know that it was the realisation by the directors of Alcoa of the importance of this outlet which persuaded them to invest massively in research in this field. The two most remarkable European products were the Bréguet 14 perfected in 1916, and the Junkers F13, which dates from 1919–20. Both of these aircrafts were made entirely from aluminium alloys, but it was not until after 1930 that all-metal construction took over. It is a fact that the development of the aeronautics industry has exerted a profound influence on twentieth-century scientific thinking. Furthermore, the problems posed by flight control, take-off and landing represented as many challenges as those faced by the developers of the electricity networks. Born of the requirements of maritime navigation, the radio-telegraph became the necessary companion to the development of aviation.

Pascal Griset has described a double trajectory of the "wireless telegraphy" in Europe between the wars: that of the technologies applied, and that of its uses. "Radio technology", he writes, "was born within one of the dominant technical systems of the end of the 19th century: electricity. Progressively radio moved away from this technical system towards a new

[20] R. Giannetti, p. 77 f. in this book.

[21] See the contribution of H. J. Braun/D. Edgerton in this book.

one: electronics. This movement was a cause of instability for the sector, but was also the mainspring of its dynamism", and "the growing diversity of the services offered by radio technology is the other factor of dynamism and instability"[22]. But the experience of European industry, compared to that of the United States, was particularly disappointing in this area, despite the importance of its contributions to technology. National markets were too narrow, and their institutions not adapted. European companies were unable to exploit the complementary aspects of different sectors such as components, professional equipment, domestic equipment and the mass media. The dominance of the United States in the electronics sector in 1945 is not only explained by the effects of the war, but also by the disappointing performance of Europe between the wars.

These various examples illustrate the major role played by interdependence in the evolution of technical systems. They confirm the importance of supply factors in the economic growth between the wars, but they also reveal the severity of European handicaps, which were due equally to the failures and segmentation of the markets and to the negative influence of institutional systems which had not yet adapted to the demands of new technology.

There is a striking difference between the dominant specialities of European and American companies. According to figures published by Alfred D. Chandler, in 1929 24 of the top 100 German companies, rated according to their capital value, produced and distributed products destined for an end-user, whereas in the US in 1930 this figure was 60%. The UK was the only European country comparable to the United States from this point of view[23].

The product strategies of German firms, both in the electrical engineering and chemical industries, were essentially determined by advances in technology, from which they benefited in a number of areas. This was the natural consequence of a research and innovation strategy whose origins dated back to the 1880s, if not to the 1860s. Developments in chemical science created new opportunities in all areas, including the dyeing, plastics and pharmaceutical industries. They led naturally to a strategy of product diversification, which was not restricted to German businesses. This was also the case at ICI and Rhône-Poulenc. In the field of electrical engineering diversification was the result of the complementary nature of

---

[22] P. Griset, p. 43 in this book.

[23] See A. D. Chandler, Scale and Scope. The Dynamics of Industrial Capitalism, Cambridge/Mass. 1990.

different kinds of equipment. Each manufacturer had to offer his customers a range of compatible materials making up a coherent system.

In electrical engineering as well as in the chemical industry success depended broadly on the capacity of a firm to adapt its products to user requirements. The conquest of a technological field by companies whose development relied on the control of scientific and technical knowledge in a particular area was not explained simply by their technical excellence; it was also the result of a constructive dialogue with their customers. Bernd Dornseifer writes about Zeiss: "At Zeiss' Microscope Division the matrix of innovation consisted of customers, in-house microscopy expertise which enabled the company to develop, diversify and test products, additional internal and external R&D capacity, and workshops. Customers continued to be a very important source of innovative ideas"[24].

Technical excellence does not necessarily imply product diversification according to the opportunities offered by a well-controlled scientific procedure. It can on the contrary be the fruit of specialisation in one specific product or category of product. A typical example of this are the Swedish multinationals which started up in the 1880s, each of which developed their own technological "niche"[25]. Among other examples of this are the leading French producers of aluminium, Péchiney, the Compagnie de Pont à Mousson, which prided itself on the manufacture of the best cast-iron piping in the world, and Michelin. In reality the vast majority of French companies between the wars only developed a single product line, and this was by no means exceptional. Most often these specialisation strategies were based on the control of one area of knowledge which may be considered to be of a technical rather than truly scientific nature.

3.2. Demand and Rationalisation

We have no accurate history of the use of the word and concept of rationalisation, which was the principal *leitmotiv* of technical literature between the wars. Indeed, as we have said, it corresponds to markedly different programmes.

An early form of rationalisation corresponded to the desire to introduce in Europe the methods of "mass production" which had been developed

---

[24] B. Dornseifer, p. 213 in this book.

[25] See R. Lundström, Swedish Multinational Growth before 1930, in: P. Hertner/G. Jones (eds.), Multinationals: Theory and History, Aldershot 1986, pp. 135–56.

in the United States just before the 1914 war in response to a mass market. This technical programme, which appeared in the 1920s, did not disappear in the 1930s. It was quite different in outlook from that rationalisation which, already in the 1920s, attempted to prioritise the reduction in sharply rising wage costs, or from that which in the 1930s tried to combat the failure in demand due to a reduction in costs as a whole. This latter form of rationalisation may be applied incidentally as much to areas of organisation as to technology.

Europe was strongly influenced by America in the field of mass consumption. One recalls that the leaders of the American feminist movement came to Europe to preach the liberating virtues of the new domestic economy based on the mechanisation and rationalisation of the home environment. In the car industry, the largest manufacturers confirmed their willingness to follow the American example and some of them, principally in Great Britain and Germany, launched ranges of popular cars. In reality, neither the structure nor the development of the market was favourable to the success of this type of car.

As an explanation for the German intense specialisation in capital goods, Dornseifer puts forward "the continuing disadvantages of small domestic consumer goods markets". He adds "German enterprises did undoubtedly have access to technological resources. What limited their opportunities for innovation was the absence of a domestic consumer goods market as large and dynamically growing as the American one, and the lack of a corresponding vision, capability and structure to perform focused product development and aggressive marketing in such an environment".[26] In fact, patterns of consumption remained influenced by the level of average incomes, which were considerably lower than in the United States. In addition, the markets for a large number of products were still subject to national, regional, social and cultural barriers. In a word, they remained segmented. Finally, as mentioned above, the institutional conditions of development of certain technologies such as electricity and radio created an additional handicap. These "structural" inferiorities in Europe were aggravated between the wars by the instability of the economy. The permanent threat of a failure in final demand had a profound influence on business strategy.

One of the paradoxes of the interwar period was the determination on the part of engineers and entrepreneurs in all European countries, despite these handicaps, to introduce methods of mass production, be they For-

[26] Dornseifer, p. 206 in this book.

dist or Taylorist. An example of this was the Ba'ta company in Czechoslovakia which managed to cut the manufacturing costs of its shoes to such an extent that the traditional European shoe industry was directly threatened. However, in the majority of cases the transposition of American methods was only partial and necessitated adaptations which greatly modified their original conception. In the case of the car industry it was possible to speak of a "British system of mass production", which was more respectful of a worker's control of his labour than the American system, but which was already leaning towards automation"[27]. In France, Citroën and Renault were the only manufacturers to apply the Ford programme in any thorough way, when they launched their assembly lines in 1933 and 1935. Michelin and the Paris-Orléans railway company, on the other hand, were the only ones to apply an out-and-out Taylorist model of planning, the former in its machine shops and the latter in its repair shops.

That is not to say that French industry did not make considerable efforts at rationalisation, but it was applied differently according to the needs of each sector. The Depression, far from slowing down this process, actually accelerated it. More or less complete forms of assembly line working spread into the mines, the food and beverages industry and into the railway workshops. The actual technological bases for this type of Fordist rationalisation are threefold: they rely on the electrification of the driving force, on the adoption of machine tools, specialised or universal depending on the case, most of which were imported from the United States or made under American licence, and they depend on the development of continuous production. However, the most common forms of rationalisation were closer to the Taylor than to the Ford model. They were extraordinarily diverse and mainly involved the adoption of more rigorous forms of organisation of work. Factories specialised within groups and within large firms. The lay-out of factories improved, and this cleaning-up programme was facilitated by the closure of many older factories during the 1930s. Timing and piece-work, but also budgetary control, quality control and scheduling made considerable advances. All these measures, the aim of which was the control of cost prices, involved not only economies of labour but also of capital. This phenomenon was not peculiar to France, for the attempt at "industrial rationalisation" dominates the 1930s throughout Europe, and it explains the increase in productivity analysed above.

[27] W. Lewchuk, American Technology and the British Vehicle Industry, Cambridge 1987.

The rationalization of labour in the factories constitutes but one aspect of a much more general reform of company structures, based on the diffusion of a functional model of organisation. This process began in the 1920s as a result of the trend towards concentration which occurred in many sectors, and intensified in the 1930s. An analysis of this vast movement of reorganisation is outside the scope of this paper. On the other hand, it is appropriate to examine the place held by R&D in these structural reforms.

3.3. Competitive Research

Since the 18th century, and particularly in France, research had been identified as a natural and necessary function of an industrial company, closely linked to the production process. But it was within German and American companies after 1870 that research developed as an independent activity, thanks to the creation of laboratories whose research was becoming more and more scientific in nature. In Germany before 1914, these laboratories had become quite large and were awarded sizeable budgets, which enabled them to carry out wide-ranging research programmes. British and French companies, on the other hand, while not ignoring the necessity for developing a coherent research strategy, did not adopt such radical measures.

Nevertheless the stake in research policies was vital, most particularly for companies which had embarked upon the path of new technology, and their importance was further confirmed in the period between the wars. The following excerpt, written in 1936, which is taken from the ICI Dyestuffs Group, is an illustration of this: "Our main problem is the highly competitive character of research work in organic chemicals. The IG believe so whole-heartedly in this field that their research effort is preponderating in this field. This is indicated by the patents taken out by the IG, almost three-quarters of which are in fields in which the Dyestuffs Group is interested" [...] "Painful evidence of IG's strength was furnished by the frequent experience of finding that when we (Dyestuffs Group) do succeed in opening up a new line of work, the IG are already there, setting up the inevitable patent barrier"[28].

"Competitive research" policies thus defined have the aim of developing a company's range of activities, and of developing and defining its

---

[28] W. J. Reader, Imperial Chemical Industries, A History, 2 volumes, Oxford 1975. See Vol. 2, p. 34.

patents. In the electrical industry, as in the chemical industry, many agreements were concluded in the interwar period based on the exchange of patents. One of the most important was the agreement signed by ICI and Du Pont in 1929, in which IG Farben refused to participate. These agreements underline the strategic importance of integrated research to firms' survival and development. The figures for foreign patents taken out in the United States confirms their effectiveness: in 1913, 34% were of German origin, 23% British and 8% French. After the war Germany rapidly re-established her position and during the 1920s confirmed her supremacy in the crucial sectors of organic chemistry and electrical equipment. In 1937 German patents represented 38%, British patents 22,7% and French patents 9% of foreign patents registered. Compared with the other two countries, a much larger share of the German patents came from large companies, which was a natural result of their research effort. It is clear that Germany owed her pre-eminent position to this policy of research.

Little is still known about the research strategies pursued by companies. Two further, complementary steps are needed to perform this analysis. We need to attempt to measure the extent of the research performed and to assess the functioning and efficiency of the system.

3.4. Extent of the Research Effort

In an article published in 1984, David C. Mowery compared certain quantitative data measuring the research effort in American and British companies[29]. He concluded that in terms of jobs the intensity of research was four to five times less in the UK, and in terms of expenditure it was three time less. He explained the inferiority of the British research effort in two ways: a less marked trend towards concentration, and an incomplete rationalisation of company structures. Since the large British companies had not been able either to accomplish their "managerial revolution" or to adopt a system of divisional organisation, they had not developed autonomous and powerful research departments.

Mowery's assessment needs considerable qualification. First of all, one of the specific traits of British research was the important role played by cooperative research. This achieved some spectacular results. Let us

[29] See D. C. Mowery, Firm Structure, Government Policy and the Organisation of Industrial Research: Great Britain and the United States, 1900–1950, in: Business History Review 58, 1984, pp. 504–31.

remember that it was in the laboratories of Calico Printers that Terylene was perfected in 1941[30]. But, above all, it is clear today that Mowery's figures need to be revised. Sally Horrocks, for instance, has pointed up the role played by basic scientific research at Cadbury's and in the food industry in general[31]. Using unpublished sources, she demonstrates that the figures achieved were far from negligible and that official estimates are much too low. Furthermore, in industry as a whole, "the majority of research [is] concentrated within a very small number of companies in a few sectors". The role of GEC in the electrical industry is dominant and ICI's spending on R&D represented between 20 and 25% of total R&D expenditure by British industry on the eve of the war.

This expenditure was equal to 2.8% of that company's turnover, as against 2.4% for Du Pont. On the other hand, a comparison with IG Farben is to ICI's disadvantage. In 1926 IG Farben had 25 large laboratories, employing more than 3,700 people. Spending on R&D reached 10% of turnover in 1926–29 and 6% in 1934–35. The annual expenditure by ICI never exceeded one million pounds, whereas from 1930 to 1938 average spending at IG Farben was 4.8 million. The figures for IG Farben are in fact exceptional "both historically and by contemporary international standards compared with the likes of Du Pont and ICI, IG Farben's main foreign competitors"[32]. One should not therefore draw any general conclusions from this comparison.

A historical reassessment is also necessary in the case of France. There are three observations to be made here, which apply beyond the borders of that country: French industrial culture has, since the 18th century, been entirely based on the establishment of the scientific nature of technical knowledge. Very early on, one particular form of research was included in the activities of numerous companies, which relied on a dialogue with the academic world and on the daily actions of engineers. The emergence during the last thirty years of the 19th century of a science-based steel industry, which is particularly well illustrated by the work of Floris Osmond and Louis de Chatelier, is quite characteristic from this point of view. This tradition persisted in the interwar period.

After 1900 and increasingly during the 1920s, research laboratories began to appear in the electrical engineering and chemical sectors. These

---

[30] See J. A. Allen, Studies in Innovation in the Steel and Chemical Industry, Manchester 1967.

[31] See the contribution of S. Horrocks in this book.

[32] Reader, Imperial Chemical Industries, Vol. 2, p. 34.

laboratories may have been modest in size compared to those of IG Farben, but some of them played an essential role in company development. This was the case at Rhône Poulenc, Saint Gobain and Péchiney, who set up or developed laboratories located at or near their factories during the 1920s and 1930s. At Rhône Poulenc the laboratory of industrial and macromolecular chemistry was part of the factory at Saint Fons, near Lyons, and the pharmaceutical laboratory was situated at Vitry-sur-Seine, near Paris, relative to the activities of the two factories near to which they were set up. The Vitry laboratory was completely transformed between 1930 and 1935. Saint Gobain had two laboratories: a central laboratory and an industrial laboratory at Aubervilliers, in the Paris suburbs. At Péchiney, factory laboratories were set up even before the 1914 war, and fulfilled the function of manufacturing control. But they gradually began to participate in research designed to improve processes and products. In the electrical engineering sector, even the highly specialised medium-sized companies such as the two Grenoble-based companies Neyrpic, which specialised in hydraulic turbines, and Merlin Gérin, founded in 1920 and specialised in circuit breakers, opened their own quality control laboratories which rapidly turned into research laboratories.

The difference in size compared with the large German laboratories remains impressive, it is true. No French laboratory employed more than a few dozen people and no French chemical company devoted more than 1% of its turnover to R&D. However, during the 1930s the directors of large companies began to have a change of outlook, despite the budgetary restrictions imposed by the Depression. They nearly all recognised that an increase in research effort was a necessary precondition for the further development, if not to the survival, of their companies. This intention became an essential element of the programmes of structural rationalisation adopted at this time. The realisation of the importance of research was not unique to France.

Nevertheless, European managers were also aware that an increase in spending on research was not enough to ensure success. The organisation and planning of research also had to be efficient. From this point of view the German experience was by far the richest and may be used as a model.

3.5. Conditions for Success

The system of research adopted by IG Farben, a world leader in the field of organic chemistry, is of particular interest. Despite the size that this company became, its managers did not adopt a truly multi-divisional system of organisation, comparable to that of Du Pont. They retained a system close to what one could describe as the European model. Research activity at IG Farben, even after the merger, although coordinated, remained extremely decentralised and closely linked to the activities of production and promotion of products. IG Farben, unlike Du Pont, never had a central research laboratory. The main decision-making bodies were the specialised "technical committees" covering the main areas of research. These were divided along lines that were constantly being changed. The committees worked in close cooperation with one another, and their activities were supervised and coordinated by a central committee with whom they established permanent contacts. Dornseifer describes this as a process of "cross-fertilisation". In addition, there were close links between research and production activities within the different divisions. The aim of this type of organisation was to avoid the growth of barriers between the research and production divisions. This was made possible by the absence of a truly multi-divisional system of organisation.

The organisation of research at IG Farben tended to preserve, within this huge organisation, the innovation model which characterised traditional European companies. Its basic principle was the integration of research into production activity. It was complemented, after the merger, by the creation of central departments whose job was to coordinate as much as to manage both activities.

In the French case, the lack of development of research laboratories in no way implies that companies were unaware of the need for a massive research effort. Saint Gobain and Péchiney are prime examples of this: as mentioned above, research was carried out mainly in small laboratories. But these were but one part of a plan which oriented the whole company towards innovative activity. Leroux Calas has shown that at Péchiney research on electrolysis tanks was quickly "transférée à l'usine" (transfered to the factory) in order to carry out full-size experiments[33]. Alcoa's solution to this problem on the other hand, was to set up large laboratories fitted with suitable equipment. But the French solution was only possible because the scientific training of their engineers was sufficient to enable them to run the two activities of research and production side by side.

The success achieved by Saint Gobain in the field of plate glass is likewise the result of a close and well organised cooperation between management, laboratory and factory. The latter were not simply manufac-

[33] See Leroux Calas, La recherche au service de la production d'aluminium, in: Histoire technique de la production d'aluminium, Grenoble 1991, p. 285–307.

turing establishments, but may also be likened to vast laboratories oriented towards R&D. New processes could be tried and tested there. The dominant characteristic, on the whole, remained the absence of separation between research and manufacturing activities. Each engineer, each factory manager had to prove his ability to improve processes and products. He could do this either on his own initiative or because the management had given him a particular task. Innovation management in electrical engineering companies was not different in conception from that in the chemical companies. The main role belonged to the research departments which worked in close cooperation with the laboratories and production workshops. German organisation, subject to the same constraints, conformed to an identical pattern.

From the beginning of the 1920s onwards there was a strong belief in the need to create central divisions capable of coordinating research, and both Saint Gobain and Péchiney set up their own divisions. A further, more decisive step was taken in the 1930s with the integration of research activities into the general reorganisation schemes. Saint Gobain's research division was awarded better financial and material resources, thanks to the development of a central laboratory. Its tasks were also more clearly defined and its authority extended. Research became one of the company's strategic priorities. Identical developments took place in the electrical engineering sector, as illustrated by Merlin Gérin, whose technical management was completely reorganised in 1937 around four dedicated research units. This reform acccompanied an administrative restructuring based on the adoption of the functional model.

It seems to us that the following conclusions emerge from our analysis:

1. There was strong industrial growth in Europe between the wars, despite monetary instability and chronic shortage in demand. This growth was due to the blossoming and diffusion of techniques combined with new investment, and, perhaps even more, to the process of renewal. These techniques were labour-saving, but also capital-saving, which is true of both the 1920s and the 1930s.
2. The evolution in the structures of the working population was characterised by two contradictory tendencies: on the one hand an industrial growth, which mobilised labour, but was insufficient to absorb the surplus in disposable agricultural labour because of its irregularity and the strong increase in labour productivity. On the other hand, there was a trend towards increased hidden unemployment in the non-agricultural sectors and particularly in the service industries. The history

of unemployment is inseparable from that of structural change and must take into account the imperfections of the labour market, only partly linked to hidden unemployment.

3. The development of the technical system was the result of the interdependence between the different fields of technology, but also of a determined and continuous effort to rationalise and reduce cost prices. In any event, the rise of European technology was hindered by the uncertainty and narrowness of its markets and by obstacles of an institutional nature.
4. European industry is characterised by the dominance of a technical culture which rejects an over-systematic separation of research activity from activities of production and product design. It prefers "in situ" research to centralised research; it is based on a coordination between research activities and those of other departments, which, it is true, can lead to the complete subordination of the former to the latter.
5. The structural reforms which took place towards the end of the 1920s and in the 1930s greatly affected research. But the attempt to centralise research remained, for this very reason, limited to the coordination function. It was only after the Second World War that European companies, or at least some of them, let themselves be tempted by the American model.

I.
The Extension of Technical Systems

Introduction

PETER TEMIN

These three papers – Giannetti on Italian power, Griset on French and English wireless communications, and Schröter on German Telephones – provide us with an introduction to the progress of new industries in the major European countries between the two World Wars. They focus on the interaction of technology and national policies. Their common theme is the way political policies affect the march of technology.

The papers are interesting both individually and collectively. Taken separately, they provide a wealth of historical detail about a variety of high-tech industries in a variety of countries. Taken collectively, they suggest important hypotheses and questions about the way these industries develop.

Giannetti chronicles the growth of electric power generation and transmission in Italy. Favored by her hilly terrain, Italy was at the European forefront of hydraulically generated electric power. The story begins with many independent generators and largely self-sufficient factories. Electric power appeared in Italy in the guise of previous power sources. Each factory had its own water wheel or turbine, its own steam engine, under previous technologies. It was only natural for each plant to continue to generate its own electric power.

This autonomous pattern suited the previous technology. Fuel – that is, coal – was easy to transport. But steam generated power was very expensive to transmit by belts or gears. It made sense therefore to locate the power source as near as possible to the consuming machines. Electric generators had the opposite costs. The motive force – water power – was exceedingly difficult to transport. And electric power could be transmitted relatively cheaply. Changing relative costs produced in time a changed pattern of power generation.

Costs militated more and more centralization of electric power generation. But there were two impediments to this process. First, the isolated power sources were configured to different standards. Giannetti cites the unusual Italian preference for tri-phase technology, differing periodicities of the generated power, and regional animosities, as barriers to integration. The result was a series of local power grids, not a national one. Giannetti laments in particular the lack of a power "backbone" running down the length of Italy.

The argument is unobjectionable as far as it goes. It would be nice to go a step further and ask how much was lost in the absence of the "backbone." Economies of scale made some centralized generation efficient. But it is not clear that the regional power grids did not exhaust the prevailing economies of scale. Even if there were still unexploited economies of scale, it is possible that the gains from a national as opposed to a regional power grid would have been very small. The issues in that case may have been largely distributional rather than technological.

Griset describes the growth of wireless communication. Noting that the name indicates only what this form of interaction is not, he describes the interwar transition from electric to electronic radio transmission and reception. Like wireline transmission, wireless could be used for point-to-point communication. And unlike wireline communication (before the age of cable TV), it could be used for broadcasting.

The birth of broadcasting created immediate issues of public policy. Recognizing at least dimly the importance of this nascent communication channel, governments wanted to keep control. They were opposed by technologically aggressive entrepreneurs that wanted to derive private benefit from their new innovations. Griset records that the struggle was a continuing one, resulting in uneasy compromises that differed between countries and shifted over time in several countries.

The private industry structure quickly reduced to a stable oligopoly. The Marconi, CSF, Telefunken companies in Europe vied with the American RCA for a world-wide market. The contest was only partly like a traditional oligopoly. The uneasy balance between public and private control was a characteristic of many countries, and the companies had to compete for political favors as often as for a technical edge.

Griset laments the untidy nature of the compromises and industry divisions that resulted from the political and economic struggles he narrates. But the alternative to this messy process is not clear. Only in America with its unfettered love of free enterprise and the accident of AT&T's private status was RCA unabashedly private. Even there, government licensing of radio stations made the process of diffusion as much political as economic. Ideologically pure solutions are not obvious.

In fact, totally private development of this valuable communications channel was unthinkable. No European country chose this option, and no other countries as well. The only pure alternative then was government domination. It would be instructive to know how the Fascist regimes in Germany and Italy handled radio transmission. While not starting from scratch, their actions still would indicate how total government control would affect the development of this industry.

Schröter summarizes the ambivalent position Germany held in the development of European telephone service in the 1920s. Politically isolated and in conflict with her neighbours, Germany was nevertheless a vital part of emerging European telecommunications. The Germans were invited to meetings on technical standards because they and the Americans were in the technological vanguard. And Germany sat in the center of Europe making connections for many other countries that wished to communication through Germany by means of German telephone cables.

The twin themes of rivalry and cooperation were played out both politically and technically. The story of telephone communication has many of the characteristics noted by Griset for wireless communication. Schröter's focus on Germany lacks the comparative quality of Griset's paper, but it provides a richly textured picture of the political struggles. For technical issues in interwar Germany were never purely internal. As described here, they always had international repercussions.

Schröter notes also that differing technical standards impeded international telephonic integration in much the same way that Giannetti shows that different technical standards impeded interregional integration of the power grid in Italy. Just as populations in isolation develop their own dialects and then languages, so too technical systems develop along separate paths when not in communication. One task in the growth of these new industries was to overcome the barriers created by the preceding independent histories.

This account of the separate papers has foreshadowed the collective theme they share. For all of the authors seek to generalize their story by appeal to a theoretical scheme for the development of technology. Two of the papers rely on Thomas Hughes' model of Large Technical Systems (LTS); Griset takes his cue from Dosi and Orsenigo. How well do these universal schemes fit the particular stories in these papers?

Despite many allusions to the wider contexts in which innovations take place, the abstractions share a focus on the technology of the system. Hughes, for example, talks of "reverse salients," "load factors," and "momentum." Dosi and Orsenigo speak of "technological paradigms." These terms describe the logic of the technology being introduced. They do not refer to characteristics of business firms or government policy-making bodies. Even though business enterprises are more prominent in the abstract discussions than political bodies, neither shares the spotlight with technology. The abstractions therefore lead writers into an undeclared technological determinism.

The papers discussed here, taken together, comprise an argument against that construction of progress. Even though the authors all take their cue from schemes that focus attention on the technology, their accounts give at least equal place to the economic and political aspects of industrial development. None of them, to be sure, argues that technology cannot be understood without reference to its economic and political aspects.

Griset therefore keeps our attention on the inability of British, French, and German governments to draw a clear line between public and private activity. Giannetti explains that the Italian state was too weak to overcome the problems caused by the differing frequency standards adopted by electric companies in different regions of Italy. And Schröter explains that state policy dictated far more attention to long-distance and international telephone service than the traffic warranted.

In each case, national and commercial interests were at least as important as technological. These new technologies were being adopted and combined into systems during the interwar years, and the process of integration and the shape of the resultant network were products of economic and political factors as well as technological ones. I suggest that this lesson of these papers is more general than the cases discussed. In almost all cases – and certainly in Europe between the Wars – political and economic factors cannot be ignored in any discussion of technology.

Innovation and Radio Industry in Europe during the Interwar Period

PASCAL GRISET

Wireless telegraphy was, at the beginning of the 20th century, a radical innovation. The break with others means of telecommunications was total from a technological point of view. Nevertheless, the innovation had to find users. The first market for wireless telegraphy was "ship to ship" and "ship to shore" communications. The tragic wreck of the Titanic in 1912 revealed to the general public the importance of the new technology. During the First World War, radio grew up and became adult. Its use on the battlefield confirmed its impact and allowed manufacturers to develop production on an industrial scale. The other consequence of this conflict was the use of radio for long distance telegraphy, mainly between America and Europe. As cables were overloaded or placed under the threat of cutting by German U-Boots, wireless offered an alternative. In 1917 and 1918 long wave stations proved their ability to pass important traffic and appeared for the first time as a potential competitor to cables. After the peace a completely unforeseen use of radio emerged with broadcasting. The first "Air concerts" occured in the United States, but the same kind of experiment happened very quickly in Europe. Then, at the beginning of the twenties, two different schemes of development emerged for the future of radio: intercontinental telecommunications and broadcasting. In these two fields radio companies had to act as manufacturers and as operators.

Analysing the evolution of this young industry during the interwar period means dealing with different kind of information. In common with other industries, radio needs to be analysed in terms of both economics and technology. But its impact on the balance of power between nations, its growing role in political expression, and its sociological consequences mean that a wider approach has to be adopted. In order to study the evolving structure of the economy and the patterns of regulation of a system, G. Dosi and L. Orsegino underline the key importance of technology. To understand the evolution of the economy: "[...] one must look at "desequilibrium" behaviours. Conversely, if one wants to understand the aggregate ("macro") order which appears in certain historical circumstances, one must look at institutional and technological structures which constrain and shape the underlying "micro" evolutionary process"[1].

This point of view fits well with the instability of the European economy between 1919 and 1939 and suggests the adoption of three fundamental "sub-systems': institutional conditions, technological regime and economic machine in order to structure this article.

1. *The Institutional Conditions*

State monopoly of telecommunications was the norm in Europe, and this extended to radio in the main European countries. After the First World War governments were faced with a new problem: the tremendous growth of radio. How were they to organize and control a new activity, which had strategic implications for the security of the nations? How could they at the same time give enough freedom to companies to allow them to gain a strong position in the international market? Two fields of activity were chiefly concerned in this evolution, international communications and broadcasting.

1.1. International Communications

Until the First World War, intercontinental communications were only a field of experiment for advanced radio transmitters. The success of the transatlantic radio link during the conflict proved the capacity of radio to convey important traffic reliably. Private companies were quickly ready to

---

[1] G. Dosi/L. Orsenigo, Coordination and transformation: an overview of structures behaviours and change in evolutionary environments, in: G. Dosi/C. Freeman/R. Nelson/ G. Silverberg/L. Soete (eds.), Technical Change and Economic Theory, London 1988, pp. 13–38, p. 28. See also N. Rosenberg, Economic developement and the transfer of technology: Some historical perspectives, in: Technology and Culture, 11, 1970, pp. 550–75. N. Rosenberg, Inside the black box. Technology and Economics, Cambridge 1982. T. Hughes, The order of the technological world, in: History of Technology, 5, 1980, pp. 1–16. F. Caron, Le résistible déclin des sociétés industrielles, Paris 1985.

organize this new service. Nevertheless, the institutional framework in which radio grew was different in France, Germany and Great Britain.

In Germany the relationship between the State and Telefunken (the private operator), seemed clear and was integrated into a coherent plan of development. D. Headrick underlines the crucial impact of this: "wholehearted backing of Telefunken by the German government"[2]. The Transradio AG, a subsidiary of Telefunken, was licensed by the German government to operate wireless stations for communications with foreign countries. The Transradio Company concession began in 1921 and extended over 30 years. Receipts from wireless messages were paid into the treasury of the company up to a dividend of 7%. Any surplus had to be paid into the German treasury. The government controlled the rates charged by the company. This legislation was not really an advantage for Telefunken, but the rule was clear and fixed.

In Great Britain and France, on the contrary, the relationship between the State and the private operators (Marconi and Radio France) were not clear at all. In Britain the Post Office had enjoyed total control over radio since 1904[3]. In this context, the old feud between the General Post Office and Marconi's Wireless constituted a real obstacle to the development of the British intercontinental network. During the interwar period the British government tried to follow too many contradictory policies: to preserve its cable network technically outdated by radio, to satisfy the ambition of the Post Office and to establish a radio-network connecting London with the Empire and foreign countries. The result was a hesitency which did not allow Marconi to maintain its dominant position over other European companies and above the Radio Corporation of America. The imperial network constitutes an example of the negative impact of the administration on the development of radio. In May 1919, the Marconi company offered to build stations and to operate a direct radio service between England and Australia. This proposal, very well received in Australia, was refused, and the Post Office decided to build its own station. In order to avoid royalties to the Marconi Company, the stations were to be equipped with obsolete technology[4]. In 1923, the new Conservative Government decided to permit private capital to enter the field of imperial

---

[2] D. Headrick, The invisible weapon, Oxford 1991, p. 185. On French attitude see P. Griset, L'Etat et les télécommunications internationales au début du XX° siècle: un monopole stérile, in: Histoire, Economie, Société, 2° trimestre 1987, pp. 181–209.

[3] Wireless Telegraphy Act.

[4] Poulsen arcs. At the same time all major international companies were equipped with high frequency alternators. Trying to justify this attitude, the Committee in charge of the problem declared: "Any good working electrician can run an arc; a skilled engineer must be

communications and to allow competition in the service[5]. A contract was finally signed in July 1924. This was severe and risky to the Marconi Company and constituted a real handicap for its future development. The private company also had problems extending its foreign network. Marconi was not officialy allowed to build and operate transcontinental stations to communicate with foreign countries before 1921. After this date the Post Office remained a competitor, establishing communications not only with colonies but also with foreign countries. The delays and limits placed by the State on the action of Marconi was a real handicap for the company.

In France the situation was almost the same. Since the beginning of the century, radio had been placed under a State monopoly. Before the First World War no exception to this law was accepted. In a climate of absurd competition between weak administrations, radio stayed underdeveloped. After the Armistice the Compagnie Générale de Télégraphie sans fil (CSF), proposed to built and operate an international radio network. Wishing to develop such a network, but unable to do, the PTT prefered to delay their answer. An agreement was finally signed in 1920 between CSF and the French administration. To develop its project, the company had to create a new subsidiary "Radio France". This new company received a thirty year concession to develop international radio communications. Administrative control was very heavy in order to prevent any "unfair" profit. A specific authorisation was needed to open each new connection. All the equipment was supposed to be transfered to the administration at the end of the concession without any financial compensation. Therefore, after a period of heavy investment during the early 1920s, the company decided to limit the modernisation of its transmitters.

1.2. Broadcasting

Europe was globally a region where regulation was heavy[6]. In Germany the monopoly tradition was very strong. In 1922 the government declared

in charge of a high frequency alternator." Imperial Wireless Telegraphy Committee; Report to Parliament, June 1920 p. 7.

[5] "In view of developments in the science of wireless telegraphy and other circumstances which have arisen since the late government decided upon the policy of a state-operated wireless chain, it is not considered necessary any longer to exlude private enterprise from participation in wireless telegraphy within the Empire." B. Law, Prime Minister, House of Commons, March 1923.

[6] See L. D. Batson, Radio markets of the world, Washington 1932, p. 14–15.

broadcasting a state monopoly on the basis that it was a form of wireless telegraphy and that telegraphing and telephoning, either by wire or wireless were government property by law. As the advocates of unrestricted radio development protested, a compromise had to be found. New companies were created. 51% of their capital was contributed by the administration and 49% privately. The broadcasting plants were constructed by the Reichpost, the companies paying a monthly fee for their use. The Reichpost was required to enlarge the plants and to install the latest improvements in radio broadcasting apparatus. In 1925, 9 broadcasting companies operated 9 main stations and 5 sub-stations. Listeners had to pay 2 marks each month. Of this amount 1,20 marks were credited to the broadcasting station; the other 80 pfennigs were retained by the Reichpost. This system remained unchanged until the beginning of the 1930s. After initial reform by von Papen, the radio fell progressively under the complete control of the Nazi regime.

In Great Britain the government hesitated between development based on private companies and that based on state owned stations. The Imperial Wireless Telegraphy Committee pronounced itself against public monopoly of broadcasting and proposed a plan of controlled competition between a public service and an oligopoly of privately owned stations. During this period of reflexion America played the role of a counterfactual repertory. Gradually, during the summer of 1922, the American system came to be identified with the "chaos of the air". This opposition to the American experience seemed to rally universal support, and monopoly came to be accepted as the only viable system. On January 18th 1923, a British Broadcasting Company was licensed. This first version of the BBC was a consortium of manufacturers[7] of domestic wireless receiving sets. Their purpose was to provide regular programmes to incite people to buy receivers. There was no pooling of patents between the manufacturers, each continued to compete for sales of receivers with the others. The system was financed through licence fees collected by the Post Office and partially reallocated to the BBC. Each listener needed a licence to be allowed to use a radio receiver. The official transformation of BBC into a public corporation in 1927 marked the end of the co-development of radio manufacturers and broadcasting. "The British Broadcasting Corpo-

[7] Besides the Marconi Company there were five other sizeable concerns with a stake in the market: Metropolitan-Vickers, Western Electric, Radio Communication Company, General Electric Company and British Thomson Houston.

ration inhereted the Company's broadcasting monopoly, its plant, its staff and its managing director John Reith"[8], but this structure was no-longer under the control of private industry. A few years after the beginning of the new media, the affirmation of the State's monopoly was total.

Compared to these two countries, the French organisation of broadcasting appeared original. In fact the term of "none organisation" would better fit the reality of the French situation. Private and state owned stations coexisted in the framework of an official state monopoly. The first radio station broadcasting regular programmes was set up by a subsidiary of CSF: La Compagnie Française de Radiophonie (CFR). After successful experiments in June and November 1921, the company had to wait until October 1922 to get authorisation to broadcast[9]. The station was called "Radio Paris". In January 1923, the PTT opened their own station "Radio PTT". This hybrid constitution of French broadcasting survived until the Second World War. This structure would have been full of potential if the administration had held a clear attitude towards private companies. On the contrary, however, during the entire period, the legislation was unable to give a real status to private stations. The first legislation concerning direct broadcasting was passed by parliament in 1923. This text was the first of a long series, with almost every change in political majority implying a modification of broadcasting policy. The permit to broadcast was "accordé à titre précaire". At any time the administration could cancel it. In such a context long term investment was impossible. Progressively, the administration's control over radio was reinforced by a profusion of legislation enacted in the 1930's (see Table 1)[10].

In France, Germany and Great Britain the institutional context appears to have been a handicap for the radio industry and was a factor which contributed to instability.

2. *The Technological Regime*

The companies involved with radio had to manage a very unstable technical situation. In common with other fields of activity, radio was technically

---

[8] T. Burns, The BBC, London 1977, p. 1.

[9] Cf. E. Girardeau, Souvenirs de longue vie, Paris 1968.

[10] Cf. P. Griset, Les révolutions de la communication, Paris 1991, pp. 67–70. P. Miquel, Histoire de la radio et de la télévision, Paris 1973. R. Duval, Histoire de la radio en France, Paris 1979.

unstable on account of the great number of innovations that regularly challenged established technologies. But, more fundamentally, the borders of this industrial field were not clearly fixed. The word "radio", used to characterize the new industry at the beginning of the twenties, hid an heterogenous reality.

Radio, as Sturmey formulates it, "[...] is not an entity, a thing in itself; it is simply the use of electromagnetic forces travelling in space. Invention in radio covers the designing of equipment for the better utilization of the properties of these forces"[11]. The evolution of the vocabulary used to denote this field of activity reveals the difficulty of outlining its territory. The first name "Wireless telegraphy" characterised the new activity in opposition to the older system of transmission. The terminology is both restrictive in term of service ("telegraphy") and very vague concerning the system ("wireless"). In order to analyse this industrial structure we have to return to the technological evolution of the sector from the beginning of the century. We will divide the elements which characterize the technology in two categories: technical and service[12]. Radio technology was born within one of the dominant technical systems[13] of the end of the 19th century: electricity. Progressively radio moved away from this technical system towards a new one: electronics. (see Figure 1) This movement was a cause of instability for the sector, but was also the mainspring of its dynamism. The growing diversity of the services offered by radio technology is the other factor of dynamism and instability. The figure gives a global vision of the interaction between these two axes of evolution and suggests three main periods.

2.1. A "Belle Époque" Technology

Between the end of the 19th century and 1906, wireless telegraphy used electrical technology exclusively. In the earliest days of wireless telegraphy

---

[11] S. Sturmey, The economic development of radio, London 1958, p. 15.

[12] Our approach is based on the concept proposed in: P. Savioti/J. S Metcalfe, A theoretical approach to the construction of technological output indicators, in: Research Policy 13, 1984, pp. 141–151.

[13] A group of technologies very widely applicable in many industries. This concept is more narrow than the concept of Système technique described by B. Gille, but it is useful to understand the technical context of the evolution of the radio industry. See C. Freeman, J. Clark and L. Soete, Unemployment and technical innovation, London 1982. C. Freeman, The economics of industrial innovation, London 1982. T. Hughes, Networks of Power: Electrification in Western Society, 1880–1930, Baltimore 1983. B. Gille, Histoire des techniques, Paris 1978.

the only method of generating radio waves was by means of sparks created by an induction coil, or sometimes a bank of capacitors[14]. A coherer was used for reception[15]. Marconi used this configuration for his first experiments. After leaving Italy, he began to demonstrate radio telegraphy over a distance of several miles in England in 1896. He formed the British Marconi company the following year[16]. This first system was only able to transmit the dots and dashes of the Morse code, but it was able to convey information. This "dirty transmitter" (spark transmitter) used with a "temperamental device" (the coherer)[17] was therefore a revolution in terms of service. During these first years radio was a craft activity, developing niche markets. Independant innovators like Ducretet[18] in France created their own company, as did Marconi. Based exclusively on the electrical Technical System radio was unable to offer a more efficient system to consumers[19].

2.2. Adolescence

Between the end of the first decade of the 20th century and 1921, a new technical system, electronics, began to offer new possibilities for radio. Spark transmitters were unable to convey voices. Both transmitters and receivers had to be improved to reach this point. Transmitters continued to be based on an electrical technology. The use of oscillating arcs generating continuous waves allowed radio to transmit speech. Poulsen[20] was the first to use an arc for radio transmission. "De Forest (USA), Colin and

---

[14] Cf. H. Aitken, Syntony and Spark: The Origins of Radio, Princeton 1976.

[15] Discovered by E. Branly in 1890. Agrégé de physique, Branly is considered as one of the numerous "father of radio". In the first radiotelegramm transmitted across the Channel Marconi evoked this role: "Mr. Marconi envoie à Mr Branly ses respectueux compliments par le télégraphe sans fil à travers la Manche, ce beau résultat étant du en partie aux remarquables travaux de Mr Branly".

[16] Marconi could be considered as the archetype of the entrepreneur définited in a model known as the Schumpeter Mark I. Marconi discovered a new idea and introduced it into economic life. He tried to enjoy a monopoly based on his patent and created his own company. He had to overcome barriers due to existing institutions. J. Schumpeter, The Theory of Economic Development, Cambridge/Mass. 1934.

[17] H. Aitken, The continuous wave. Technology and American radio, 1900–1932, Princeton 1985, p. 5 and 9.

[18] Cf. A. Vasseur, De la TSF à l'électronique, Paris 1975.

[19] Mainly Navy.

[20] Born in Denmark in 1869.

Jeance (France), Elwell and Fleming also improved arc transmitters"[21]. This technology became dominant around 1914. Another electrical technology, high frequency alternators, led to important progress, and allowed long distance transmissions. Alexanderson[22] was the first to propose a device fit for commercial use in 1908. Goldschmitt in Germany and Bethenod and Latour in France took this concept further when they designed machines for Telefunken and for the SFR[23]. This new way of generating continuous waves surpassed arcs at the end of the First World War. The activities of companies involved in commercial long distance radiocommunications were based on the use of long wave transmitters, mainly alternators[24]. These devices were expensive to build and to operate. The cost of a complete station was extremely high. 60 million Francs, for example, for the French station of Saint Assise, the most powerful in the world in 1921[25]. As well as being expensive, the alternators were also very sophisticated[26]. In the case of the French alternator Béthenod Latour, the rotor had to run in a vacuum, to avoid air friction and a pump was needed to maintain the system. The rotor and the main bearings had to be kept cool by forced oil circulation, which necessitated another pump[27]. The alternator was nevertheless the only available technology for the development of an efficient intercontinental radio network.

Developments in the field of receivers were, from a long term perspective, much more fundamental. The most important advance was the thermionic valve, with which the names of J. Fleming and Lee De Forest are particularly associated. On the basis of the Edison effect, Fleming constructed the first diode valve in 1904[28]. This new device, patented by Marconi Wireless, improved dramatically the efficency of receivers. The

---

[21] W. Dalton, The Story of Radio, 3 volumes, London 1975, Vol. 1 p. 98.

[22] Born in Sweden in 1878.

[23] Société Française Radioélectrique.

[24] Three companies controlled this technology: RCA (Alexanderson alternator), CSF (Béthenod-Latour alternator) and Telefunken (Goldschmidt alternator). Marconi had no equipment of this quality at the end of the war.

[25] Cf. E. Girardeau, Souvenirs de longue vie, Paris 1968.

[26] Cf. P. Griset, Triomphe et désuétude de l'alternateur à haute fréquence Béthenod-Latour. La France des électriciens, Paris 1986, pp. 261–71.

[27] The Imperial Wireless Telegraphy Committee recommended avoiding the use of this too complex and too expensive technology. Imperial Wireless Telegraphy Committee; Report to Parliament, June, 1920, p. 7.

[28] So called because it restricts the flow of electricity to one direction just as a valve controls the flow of water in a pipe. Cf. P. Griset, Triomphe et désuétude de l'alternateur à haute fréquence Béthenod-Latour, pp. 261–271.

decisive innovation came two years later. Lee De Forest[29] applied for a patent for a three-electrode valve which acted as a detector, and could also be used to amplify weak currents. This last characteristic constituted the basis of a new technological system: electronics. The triode: "[...] rendered many other promising inventions more or less abortive and caused many thousands of pounds worth of apparatus to be discarded"[30]. In term of service, these developments allowed the radio industry to propose reliable, but relatively expensive, intercontinental transmissions and portable short range receivers. If the impact of electronics was already significant, the heart of the system was still based primarily on the electrical technical system.

2.3. Maturity

Between 1923 and 1939 decisive developments occured in both service and technical terms. From a strictly technical point of view, the main change was the use of short wave transmitters for long distance telecommunications. Until the end of the First World War it was generally considered that the lower wavelength limit for commercial transmission was two hundred meters. Shorter wavelengths were assigned to amateurs[31], who quickly proved that short wavelengths had unsuspected possibilities. With very cheap transmitters they achieved transatlantic transmission at the end of 1921[32]. On receiving information of these results Marconi decided to erect an experimental station at Poldhu to work on the one hundred meters wavelength. These experiments were extremely successful. Speech was transmitted between England and Australia for the first time on May 30, 1924. This method, called the "Beam system", was adopted by large companies in 1926, when the use of shorter wavelengths permitted the transmission of messages both at night and during the day. By using reflectors the greater part of the wave energy was confined to a directional

[29] Cf. L. de Forest, Father of radio: The autobiography of Lee de Forest, Chicago 1950. G. Carneal, A conqueror of space: An authorized biography of the life and work of Lee de Forest, New York 1930. R. Franc, Eugène Ducretet, Editions du tambourinaire, Paris 1965.

[30] A. Morse, Radio beam and broadcast. Its story and patents, London 1925, p. 38.

[31] International Radiotelegraph Conference, London 1912.

[32] During a series of tests beginning the 22nd December, 20 English amateurs, 14 French and 6 Dutch were listing into the USA. Cf. A. Vasseur, p. 150.

path[33]. Economy of power[34], speed[35] and secrecy of transmissions were the decisive consequences of this innovation. This new generation of transmitters needed less power than longwaves, were cheaper and used valves to generate radiowaves. Although transmitters based on electrical technology remained in use for a few years, the dominant technology was henceforth electronics. The kind of service provided by radio changed radically with the emergence of broadcasting. Broadcasting was not really a technical innovation. It is more accurately classified as a social and cultural creation. In fact the technologies used for broadcasting and for point to point radio telephone were the same. But the purposes were completely different. This revolution radically transformed the radio industry. In a few months companies had to become show organizers and mass-production manufacturers. This period was marked by the maturity of the radio industry and the emancipation of electronics . Considered as a part of this sector since its origin, electronics became adult and can be considered as a real new technical system. (see Figure 2).

The development of radio can be seen as a two stage process. The first part was the emergence of a pre-technical system, characterized by a high level of empiricism and the concentration of innovators on solving operational problems arising from the incapacity of the dominant technical system to support further development of new services. When these tensions decreased, and when scientific knowledge was able to follow, and even to precede technical evolution, the pre-technical system was able to generate by itself new services or products. A new technical system was born.

3. The Economic Machine

Technological evolution had a considerable influence on the economics of the industry. The rhythm of change in technical systems seems to define a periodization in the evolution of firm structures. Consequently, the way companies and nations dealt with this technologically based instability determined their ability to conquer, or at least to defend, market shares.

---

33 Cf. A. Morse, Radio beam and broadcast. Its story and patents, London 1925.

34 Power of the transmitter divided by ten.

35 200 words a minute with beam system, 100 with long waves apparatuses.

3.1. The Market

Radio was necessary for countries like France, Germany and United States to escape from the British hegemony in transcontinental communications. "No sooner did the guns go silent on the Western Front than the old conflicts over world communications reappeared. Great Britain, still dominating the world's information flow, was once again challenged by nations that resented its power"[36]. The development of international radio communications in the years immediately following World War I has to be understood as a double stake: strategic and commercial. These two goals were inseparable, and to succeed nations needed a strong link between diplomacy and the activities of private companies.

The new balance of power emerging from World War I opened many opportunities. In Central Europe, in South America, and in China European countries had to compete to recover their influence against a new competitor: the United States. Four companies sought a strong international position: Marconi, Telefunken, CSF and RCA. The competition took different forms. The goal was to obtain the maximum number of licenses from foreign governments in order to gain future control over the most important part of the traffic, and to find an outlet for professional equipment. The organisation was usually based on a triangle composed of the manufacturer, the main operator and the local operator. The main operator was supposed to obtain a license from a foreign government to build and operate a radio station for intercontinental communication. This first step was usually the most arduous. The strong support of the embassy was needed, and a good knowledge of "local financial tradition" was also important. The German strategy was, for example, based on the activity of Telefunken (manufacturer) and on its subsidiary Transradio (main operator). When Transradio succeeded in obtaining a licence, a local subsidiary, in cooperation with local investers, was created. This new company bought equipment from Telefunken and communicated with the central station of the Transradio network, Nauen. The same system was used by CSF[37] with its subsidiary Radio France (See Figure 3 and 4).

---

[36] D. Headrick, The invisible weapon, p. 173.

[37] In 1918 the principal French radio interests were combined in a single holding, the Compagnie Générale de Télégraphie sans Fil (CSF). This company controlled Société Française Radioélectrique founded in 1910 by Emile Girardeau. The SFR was a relatively small manufacturer until the First World War. During the conflict the SFR increased dramatically its production of radio transmitters to equip French Army and Air Force. Cf. P. Griset, La Société Radio France dans l'entre deux guerres, in: Histoire, Economie, Société, 1° trimestre 1983, pp. 83–111.

European companies were highly efficient at this kind of marketing. In South America, for instance, they obtained numerous licenses from government and were in position to build many stations. Despite the Monroe Doctrine the Radio Corporation of America had to share this market on a parity basis with the three European companies. This deal was included in a wider agreement, including patent cross-licensing between the four companies. This international cartel, to be known as the AEFG Consortium revealed the strong position of European companies in the international market[38].

In the personal radio receiver market, protectionism appeared very quickly in Europe. Some radio laws included restrictions on the importation of radio equipment. In the United Kingdom this was the case in the first part of the 1920s. Even when the law was changed, the habit of a protected market remained strong. With a view to protecting British made wireless materials, the law contained a hard condition for users: "[...] the licensee shall not knowingly use any set or component part manufactured elsewhere than Great Britain, Northern Ireland, the Channel Islands or the Isle of Man"[39]. This era ended in 1924[40]. After December 24, 1924, the licensees were allowed to buy any kind of sets, without regard to their origin. This reform was perceived as unfair by British manufacturers. They emphasized that they had lain the foundations of broadcast entertainment when nobody was certain whether it would be successful or not, taking all the risks themselves. A few days before the law took effect "The Times" summed up the feeling of the manufacturers. "[...] the feeling among British makers of wireless sets and apparatus was that it was desirable for the embargo on foreign articles to remain. Without doubts, its removal would strongly encourage foreign competition and tend thereby to increase unemployment [...] for the really high-class British product

[38] On the contrary in China, no deal could be found and each company played its own card. Cf. L. Tribolet, The international aspects of electrical communications in the Pacific area, Baltimore 1929.

[39] It was very difficult to apply this text. If the control of complete sets was possible, the control of components was much more arduous. In fact foreign made parts, principally German, French and Dutch, were being sold in considerable quantities. Nevertheless, such sale was not conducted openly. No advertising of foreign material was done and precautions were observed by the dealers so that this part of their activity did not come to the attention of the National Association of Radio Manufacturers. This organisation, partially based on protectionism was clearly suited to the situation of a young industry.

[40] In order to follow the definite expression of opinion by the Broadcasting Committee of 1923 that wireless receiving licences should contain no condition respecting the origin of the apparatus used.

the demand of the initiated would remain; but the threat contained in a possible flood of cheap foreign goods was none the less serious [...]"[41].

In fact, even after the change in legislation, the English market was still controlled by local manufacturers. The National Association of Radio Manufacturers and Traders included in a single composite scheme the organisation of all three important trade groups – manufacturers, whole salers and retailers. The manufacturers' control was exercised through a regulation which meant that no wholesaler was able to become a member of the association if his application was rejected by the manufacturers group. The rules of the association stipulated that members could not deal in foreign made goods. In cases of nonfulfillment of this regulation, a member was black-listed, and no member was allowed to deal with him. The mentality of British manufacturers was a curious mixture, bringing together the need for protectionism with an apparent self confidence based on the supposed superior quality of equipment made in Britain.

With regard to competition from Germany : "[...] local dealers and users seem to be of the opinion that while German prices are far below those of the United States and England, the quality of the apparatus is proportionaly low"[42]. The analysis of the market was based on the decisive impact of quality: "There will always be a demand for the best quality apparatus, particularly by radio users, because on the quality of their apparatus depends entirely the quality of their receptions." declared a dealer reflecting the general opinion. From this point of view: "American competition [...] is even less to be feared than German competition"[43]. According to inquiries made by a representative of "The Yorkshire Observer" the majority of dealers admitted that as a result of foreign competition, parts and sets would be sold more cheap, but many of them believed that British quality would beat cheaper, but inferior, imported goods[44].

This kind of attitude was very common in Europe. The mentality of French manufacturers was almost the same. The Depression reinforced this attitude. The competition coming from the United States increased from the end of the 1920s. Prior to 1929, American participation in the European markets was not important. Confronted by a fall in demand in their domestic market, American manufacturers decided to pay more

---

41. The Times, December 18, 1924.

42 Report: New development in British Radio. F. C. Lee, American Consulate, Bradford England to The Secretary of State, December 23, 1924. US National Archives, RG 38 Box 377.

43 The Yorkshire Observer, December 22, 1924.

44 Cf. ibid.

attention to Europe. From 1930, long waves sets, different from the ones used in the United States, were specially built in American factories for European use. "Whether it is considered that the appearance, tone, quality, life, price or other factor may have supplied the reason, it is true that from the point of view of the European manufacturers, American products in the radio field were becoming alarming popular and successful"[45]. Restrictions on imports, exclusive of ordinary import duties and business taxes, increased rapidly during the fall and winter of 1931–1932. The method ordinarily employed was to select a specific period in the past and assign to each exporting country a quota based on the percentage of imports from them during such a period. This quota concerned mainly American appliances. Nevertheless the main weapon used to fight American imports was patents. In many European countries, the distributors of American sets were sued for infringement to patents. In Austria, for instance, Telefunken and Philips had established priority on the majority of radio patents. "Import permits are required for American sets [...] these permits are issued only with the approval of the recognized patent holders [...] it is impossible to obtain them"[46]. These two companies had a very strong position in Europe concerning patents. Philips was the owner of six key patents in several countries. Its position was extremely strong in the field of tubes. With Telefunken and sometimes Marconi, they organized patent pools in many European countries[47]. The European market was therefore fragmented and did not leave any opportunity for real competition between manufacturers.

3.2. Production and Structures

If European companies succeeded in making high level professional equipment, they had difficulties adapting their aproach to the demands of mass production.

The French consortium CSF was, for instance, confronted by serious difficulties. Its subsidiary, the SFR, failed to make cheap and efficient radio sets. Making and selling of the professional equipment and of that for the general public were placed under the authority of the same people

[45] L. D. Batson, Radio markets of the world, Washington 1932, p. 19.

[46] Ibid.

[47] Cf. ibid. The situation in each European country is analysed by the American commercial attaché.

and were organized in the same place. The SFR, which was very efficient in the field of long distance transmitters and military equipment had to face a new kind of problem: organizing mass production and selling products to "monsieur tout le monde"[48]. If the basic technology was the same, the organisation of the factory as well as of the marketing were totally different. The sale of "Radiola" radio sets begun in 1922, and increased significantly in 1923. In this year 6 million francs worth of equipment were sold, of which 17% were exported. The managers of the company were disappointed by these results. In November 1923 it was decided to totally reorganize the dealer networks. At the same time the product was redesigned to meet public demand, a criteria neglected until then by the engineers. Technical efficiency was not enough. The new radio sets had to be easy to use, beautiful and cheap. The clients were no-longer the amateur enthusiasts but became hard-to-please consumers. During 1925 the situation improved. The board of managers noted the growing quality of the receivers, which were smaller and equipped with a new loud speaker device recently patented by the company. Nevertheless, investment in the factory stayed at a very low level, 200,000 francs in 1925 as production grew 27 per cent. This lack of investment caused significant problems. The organisation of prodution and the mobilization of the labour force were more and more difficult. The development of radio sets was planned in the company's laboratories, but a long delay existed between the conception of a new product and its commercialization. In 1927 new models were at last proposed. Batteries were replaced by an ac/dc supply. The Radiola SFR 20 was smaller and cheaper than its predecessor, and was equiped with a new system called "Unique command" for tuning. Unfortunately the factory did not seem to be able to fulfil the new manufacturing criteria. Returning from America, where he had visited radio set manufacturers, E. Girardeau declared to the board of managers: "We have to increase our production, to reorganize our methods of making, by the application of work in quick succession by the means of more modern tools and machines." These methods, well known in other branches of French industry seemed exotic to the managers of the radio company. The same kind of syndrome seems to have occurred in the United Kingdom. "The price of British sets, in the absence of mass production methods, was too high"[49].

---

[48] Data concerning the SFR were collected in the archives of the Banque de Paris et des Pays Bas and in the Reports of the Board of Managers (Archives Thomson).

[49] Sturmey, p. 161. The advance of American manufacturers was evident. See: D. Hounshell, From the American system to mass production: The development of manufacturing technology in the United States, Baltimore 1984.

The evolution of CSF offers an example of the impact of this phenomenon on the structure of a group. During the period of constitution of the CSF group (see Figure 5), the objective was clear: to exploit all the opportunities offered by radio technology – to be present everywhere. The framework of development was strongly based on technological know-how. A few years later, at the end of the 1930s, despite the existence of the same companies, the real structure was different (see Figure 6). The strategy of development adopted in the 1920 was curtailed. A reorganisation of the group's activities occurred in order to take account of the group's financial and technological potential. Some sectors were stressed, while others were progressively abandonned. The group concentrated its activities in the field of professional equipment. On the other hand it disengaged from consumers' products and operators' activities.

The Compagnie Française de Radiophonie ceased any activity in this field in 1934 when the administration nationalized the station "Radio Paris". Although it was presented to the public as an example of the State's authoritarianism, this move was in fact beneficial for CSF.

Radio France continued to operate the intercontinental network. But, as the end of the licence was near, the group stopped investment and considered this part of its activities as secondary.

At the end of the 1930s, CSF decided to give up this sector progressively. Defeated by Philips in crucial patent suites concerning valves, CSF reorganized the making of components. La Radiotechnique was sacrified in this new configuration. All the "safe" activities (in term of patents) were transfered to SFR. This was mainly valves for professional equipment. On the other hand, La Radiotechnique retained the production of valves for consumer equipment, and received from the SFR the control of production of consumer radio sets. Progressively Philips came to control La Radiotechnique[50]. This redeployment was, in fact, a retreat from the consumer equipment market. At the end of the 1930s, CSF was nolonger a "radio group" but an electronics group concentrated on the making of professional components and equipment.

Marconi mirrored those developments closely. In 1923 their receiver business was developed by a subsidiary: Marconiphone. At the same period as CSF, Marconi gave up its activities in receivers and consumer valves. "The Marconi interest in Marconiphone was sold to the Gramophone Co Ltd in

---

[50] The control was total in 1947 but effective since the middle of the 1930'. A large part of the radio sets commercialized under the CSF's trade mark "Radiola" was in fact imported from Netherland.

1928 [...] The Marconi interest in the Marconi-Osram Valve Company was sold at about the same time"[51]. Like the French group, its British counterpart decided to concentrate on professional equipment.

4. *Conclusion*

This study of the radio industry in the interwar period allows us to conclude with three main points.

1) From the perspective of the historian of technology, this period was decisive in the genesis of a new technical system: Electronics. "Radio" has to be analysed as a "pre-technical system", a kind of intermediate step before the affirmation of electronics as a plain technical system. More globally, this model of evolution could be tested on other technical fields and eventually constitute another key of understanding for the historian of technology.

2) From an economic point of view the analysis of the development of firms points towards the destiny of European industry after the Second World War. In fact, beyond the term "radio industry" the reality of this sector shows the genesis of the major new industrial sectors of the second half of the 20th century:

Electronic components
Electronic professional equipment
Electronics for consumers
Mass media

The "historic" companies (Marconi, Telefunken, CSF) tried to manage all these activities simultaneously and mostly failed in this ambition. They were, in fact, real pioneers, discovering new applications for equipment mainly based on the triode discovered by Lee de Forest.

In order to resist competitors and to ensure the maximum development of their company, managers tried to develop all the activities linked to their technological know-how within a common structure. As the success of RCA in the United States seems to demonstrate, the failure of Marconi, CSF and Telefunken did not mean that their analysis was wrong. RCA created global control over the new sector, including professional equipment, consumer electronics and mass media. European companies were

---

51 Sturmey, p. 166.

not able to evolve in the same way. Market structure and institutional context explain, for the most part, these divergent developments. If, before the Second World War, the consequences were not evident, they appeared much more clearly after 1945. Although the dramatic impact of World War Two largely explains the dominance of the USA in the field of communications, the lack of an existing industrial basis in Europe has to be kept in mind. The two main trump cards leading to success in this field were the simultaneous control of program and receiver business (Software and Hardware) and the financial and technical link between consumer electronics and professional equipment. These were missing in Europe. This wider chronological angle also changes the meaning of our study. In fact, the interwar years cover the first period of the electronics industry and perhaps the prehistory of the information industry[52]. During these first decades European companies failed to create the conditions for success. Their concentration on the professional and military equipment was largely explained by their difficulties in managing mass production and price competition.

3) A strong link between the evolution of the technical system and the level of coherence of the industry clearly emerges. Three periods could be defined (see Figure 7):

During the early years the activity of the sector was based on a single technical system: electricity. The number of services offered to customers was small and easily manageable. Until 1925, the coherence of the companies decreased very quickly. The activity was still based on electricity but electronic devices started to become more and more important. In term of service, coherence was also decreasing. If the emergence of transcontinental communications allowed companies to conserve a minimum of coherence, the development of broadcasting created a real unsteadiness.

The difficulty of managing such heterogeneous groups and the impact of the depression forced the radio companies to reorganize their structure. This movement began in 1927–1928, when the need for research in electronics increased, and when the explosion of the radio set market made the difficulties worse. The decrease of earnings linked to the depres-

---

[52] The concept of information technology, including telecomunications and computers, is increasingly present in recent analysis. See: H. Poppel, Information technology, New York 1987. P. Jowett, The economics of information technology, London 1985. For a wider historical approach: J. Beniger, The control revolution, Cambridge/Mass. 1986. C. Marvin, When old technologies were new, New York 1988. A. Beltran/P. Griset, Histoire des techniques aux XIX° et XX° siècles, Paris 1990, chap. 4, pp. 133–82.

sion sped up this development. CSF (see Figure 8) offers an example of this attitude, but all the European companies had to make their own choice, and to reduce the range of their activities.

These conclusions are mostly based on the study of a few companies in France, Germany and Great Britain, their validity is only partial. In fact there is a lack of historical studies concerning the economies of information in Europe. The recent evolution of our societies could encourage such work. Dealing with political, cultural, economic and technological dimensions, they will have to be conceptualized on the methodological basis of global history.

Table 1: 1930's radio legislation in Europe

Broadcasting and receiving unrestricted:
Portugal.

Broadcasting unrestricted, receiving restricted:

Belgium:	(intallation fee, annual license)
Estonia:	(annual license)
Finland:	(annual license)
France:	(annual license)
Netherland:	registration no fee
Spain:	(annual license)

Broadcasting and receiving restricted:

Austria:	broadcasting monopoly, annual receiving license
Bulgaria:	broadcasting concessions, annual receiving license
Czechoslovakia:	broadcasting monopoly, annual receiving licences
Danzig:	broadcasting monopoly, annual receiving license
Denmark:	Government broadcasting monopoly, annual receiving license
Germany	Broadcasting limited to corporations using government stations, annual receiving licence.
Greece:	broadcasting prohibited, annual receiving license
Italy:	broadcasting monopoly, annual receiving license
Latvia:	broadcasting monopoly, annual receiving license
Lithuania.	broadcasting monopoly, annual receiving license
Norway:	broadcasting concessions, annual license
Poland:	broadcasting monopoly, annual receiving license
Rumania:	broadcasting monopoly, annual receiving license
Sweden:	broadcasting monopoly, installation fee, annual receiving license
United Kingdom:	broadcasting monopoly, annual receiving license

Despite the same basic legislation (State owned monopoly), the evolution of the institutional context diverged in France, Germany and Great Britain.

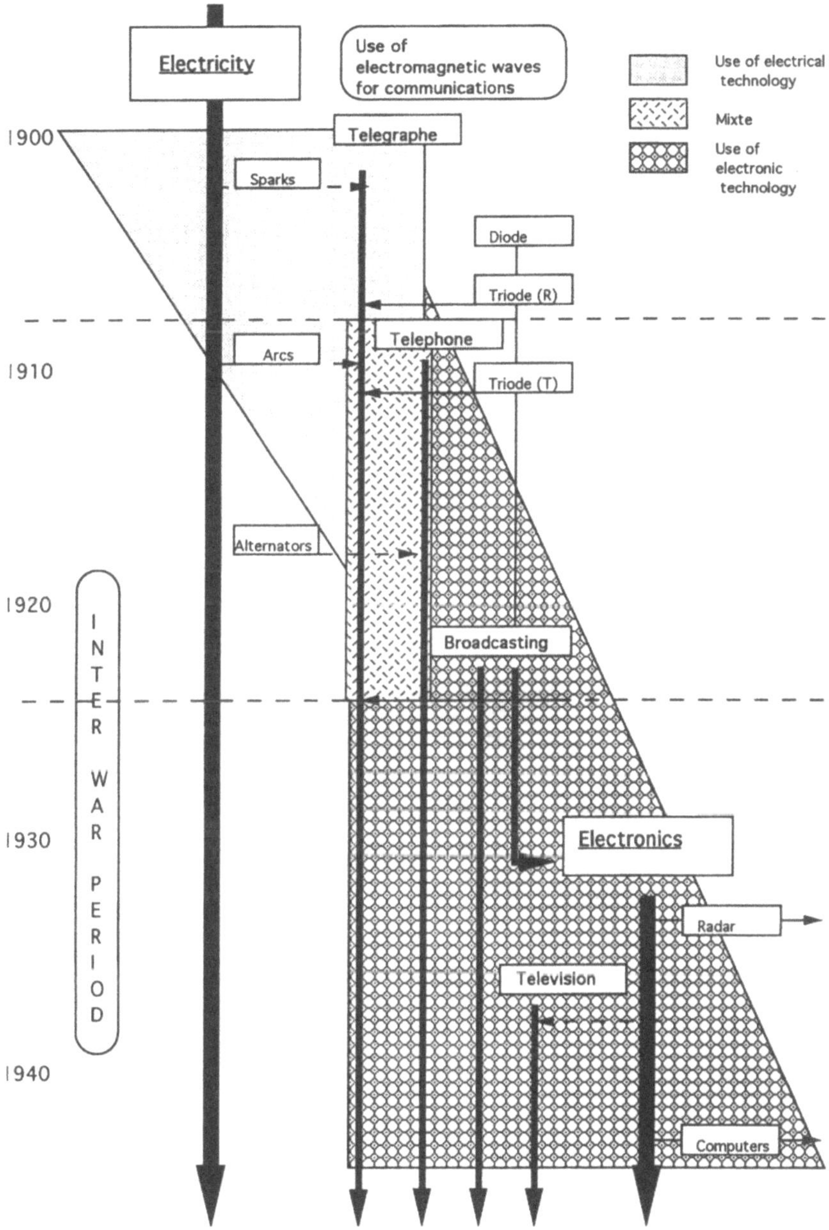

Figure 1: Use of electromagnetic waves for communications

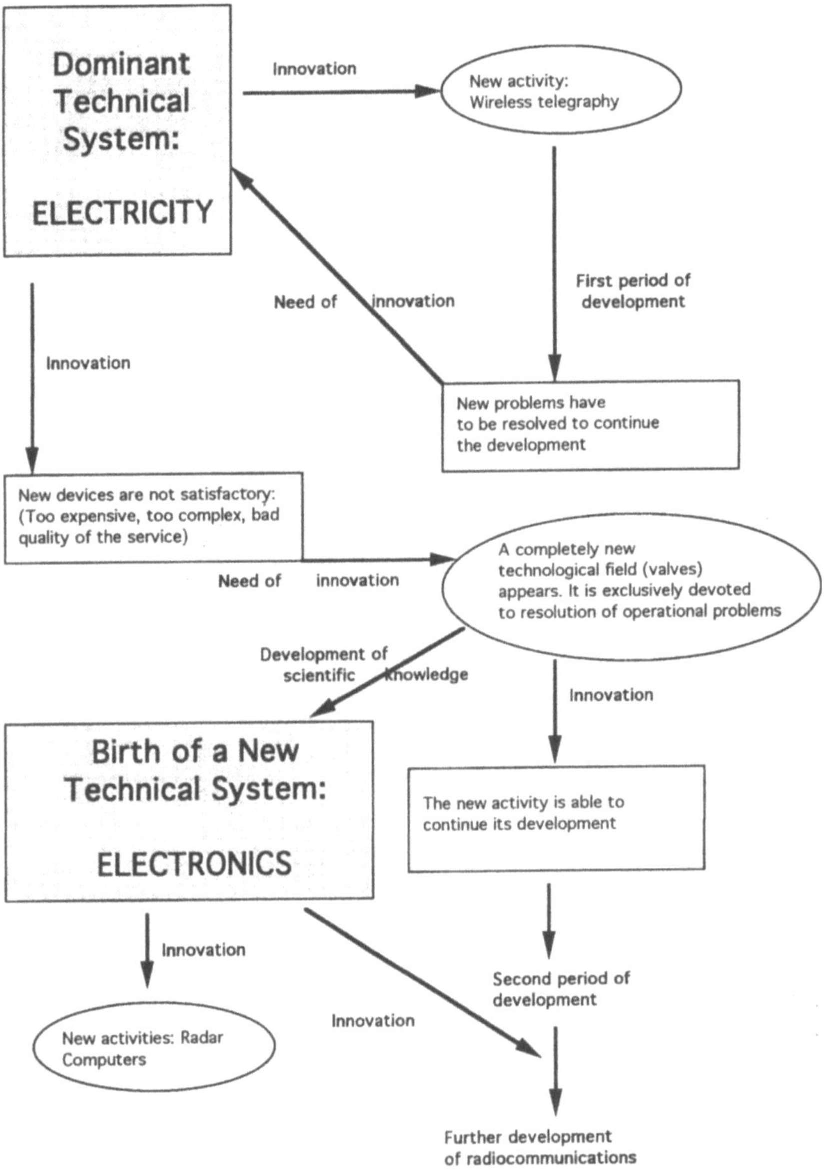

Figure 2. Electronics as a new technical system

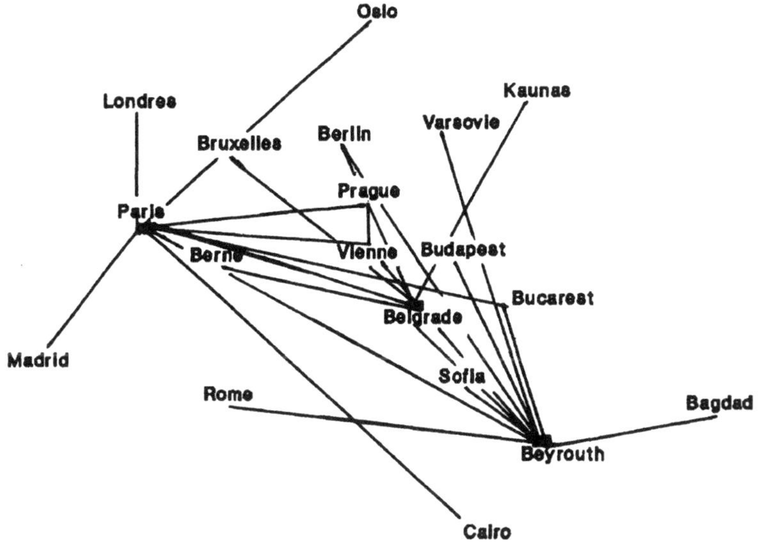

Figure 3: European Radio Network controled by CSF

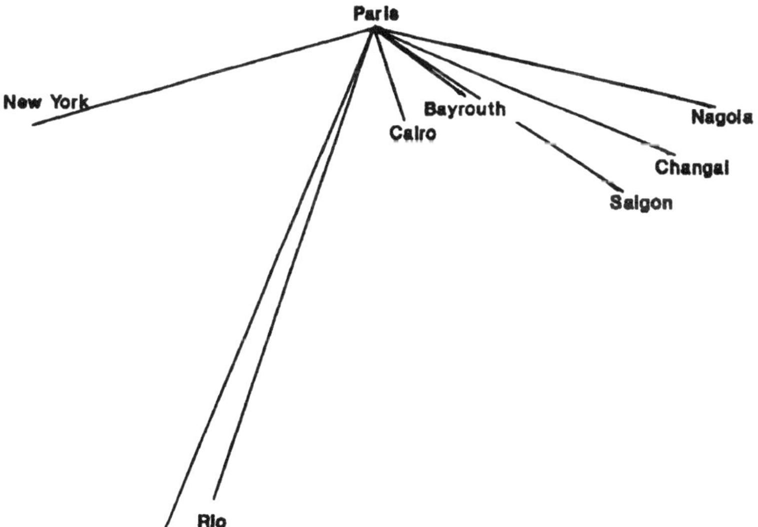

Figure 4: Intercontinental Radio Network controled by CSF

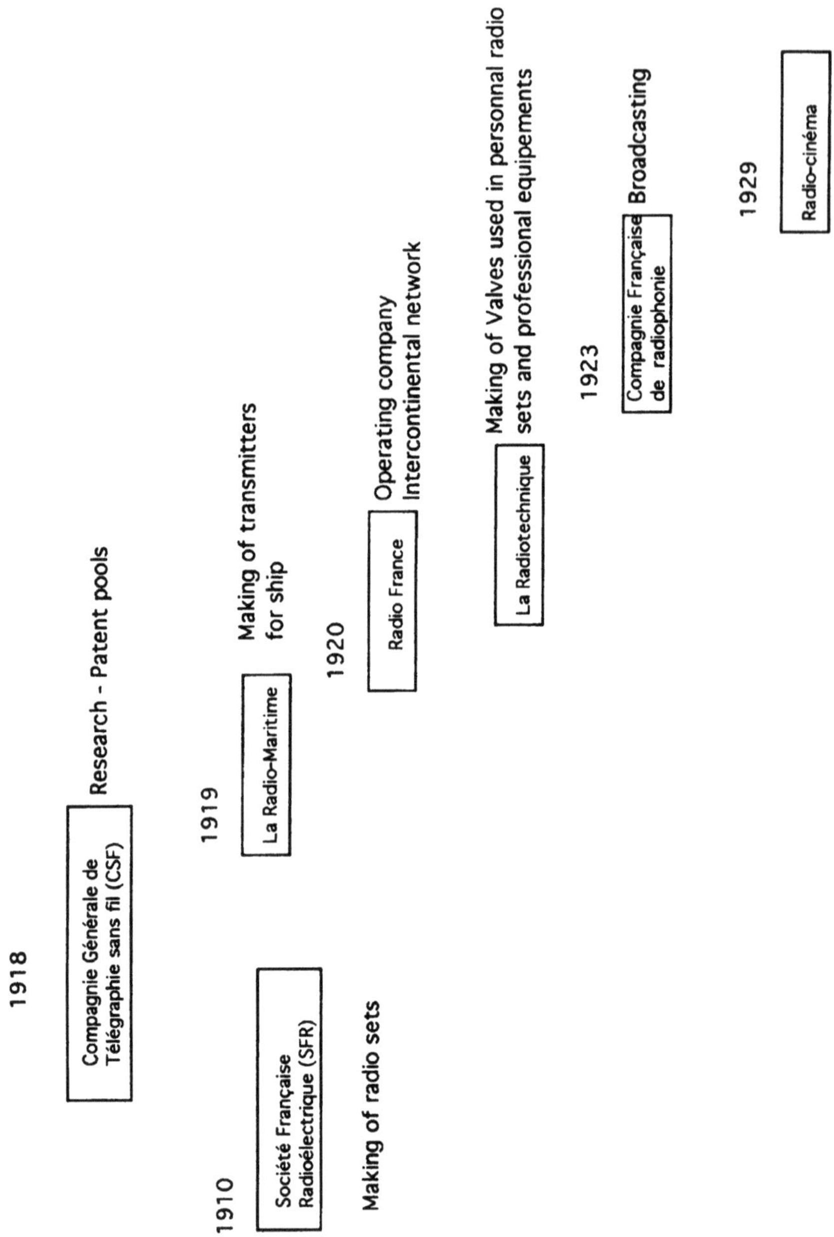
1910
Société Française Radioélectrique (SFR)
Making of radio sets
1918
Compagnie Générale de Télégraphie sans fil (CSF)
Research - Patent pools
1919
La Radio-Maritime
Making of transmitters for ship
1920
Radio France
Operating company Intercontinental network
La Radiotechnique
Making of Valves used in personnal radio sets and professional equipements
1923
Compagnie Française de radiophonie
Broadcasting
1929
Radio-cinéma

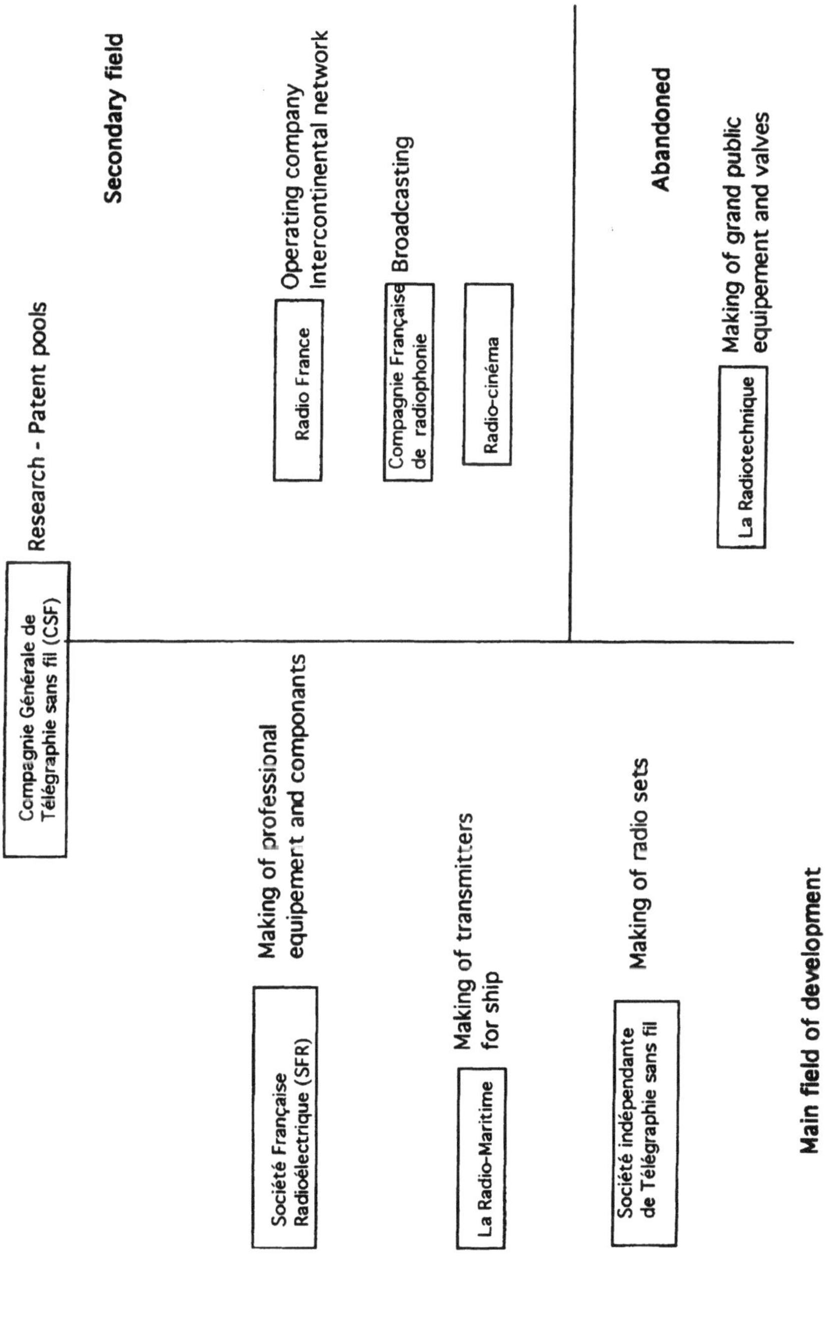

Figure 6: 1930's Organization of the CSF group

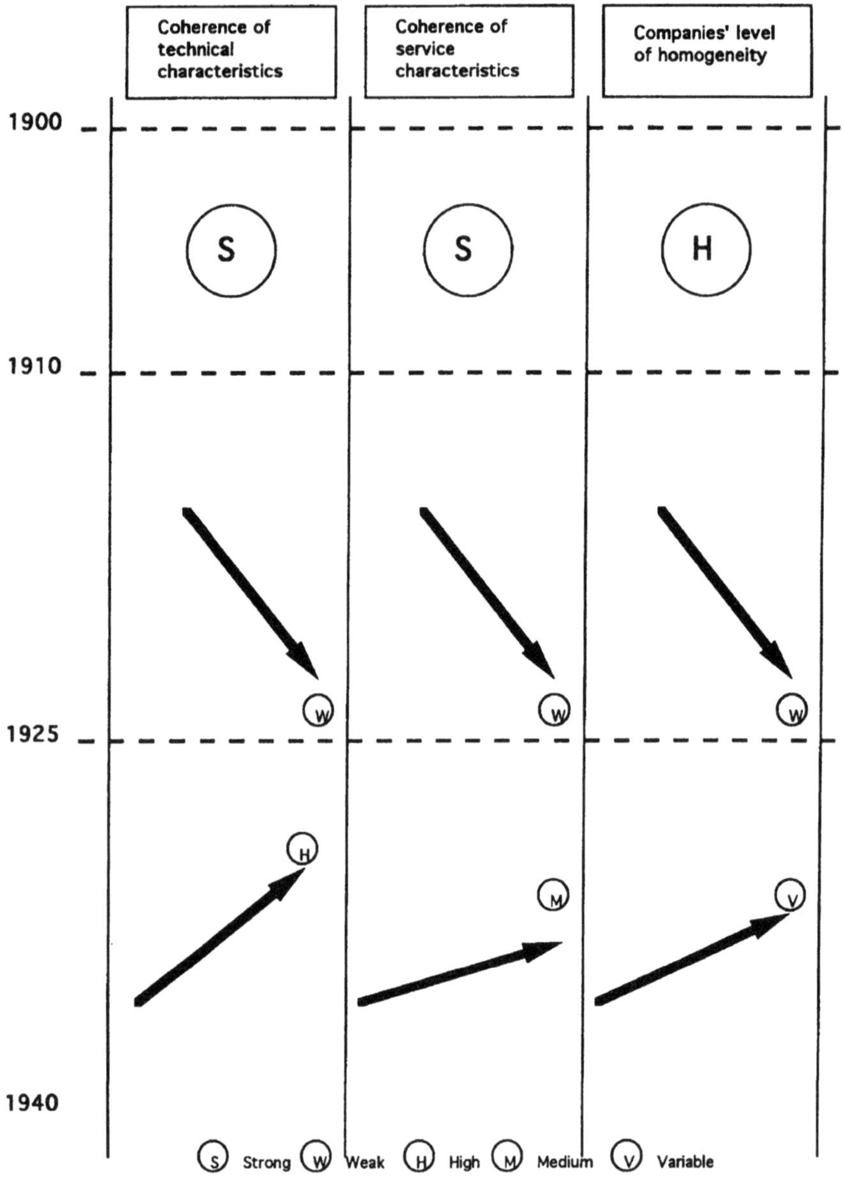

Figure 7: Evolution of the technical system and level of industry's coherence

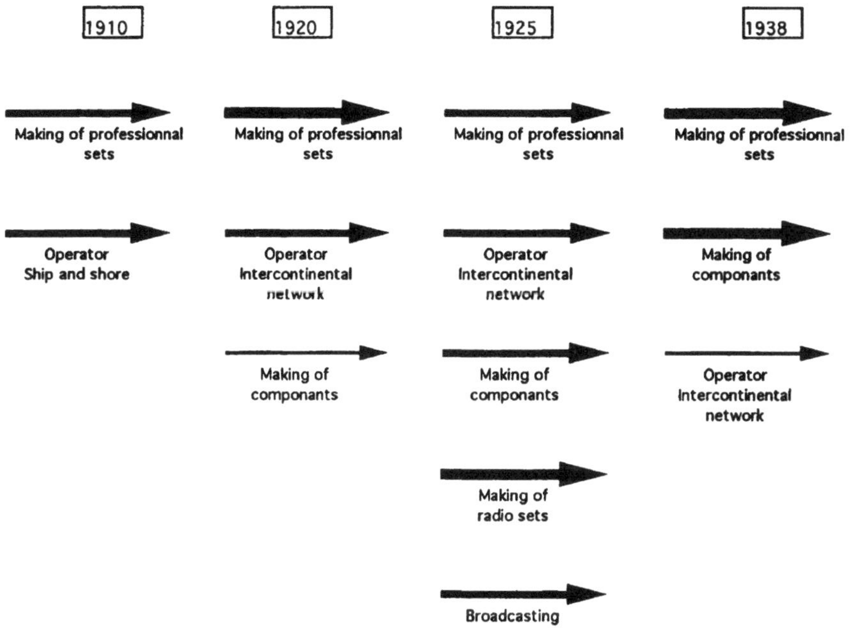

Figure 8: Business activities of CSF, 1910–38

From Small Insulated Plants to Regional Networks: The Path of Growth of the Italian Electrical Industry from its Beginning to the 1930s

RENATO GIANNETTI

An important body of the recent literature on the genesis, growth and diffusion of technical change especially that by E. Constant, G. Dosi, D. Sahal, and T. P. Hughes[1], suggests the existence of a group of basic components, technological guideposts or trajectories, which give common features and direction to all the different historical expressions of an innovation. Nevertheless, within this common "trajectory", hindsight has shown that the historical pattern of the various national experiences also depends upon localized factors, i. e. the nature of the resources used, the secondary characteristics of the technology selected, the changing social and political conditions, and the strategies of management adopted by firms, which generally ensure a close relationship between the original technological and organizational features and further innovations[2]. In this paper, I shall examine the first fifty years of the Italian experience of electrification in the light of this interpretative framework.

1. The General Features of the Italian Electrical Industry

The production of electrical energy is particularly sensitive to problems of location and to the availability of resources. The shape of an electrical system is, therefore, dependent on several factors:

---

[1] Cf. E. Constant, The origins of the Turbojet Revolution, Baltimore 1980; G. Dosi, Technical Change and Industrial Transformation, London 1983; D. Sahal, Recent Advances in the Theory of Technical Change, Berlin 1979; T. P. Hughes, Networks of Power: Electrification in Western Society, 1880–1930, Baltimore 1983.

[2] On these themes, localized technical change and, more recently, the process of path dependency and diffusion lags, cf. P. David, Path dependence: putting the past into the

1) The kinds of technology available.
2) The location of specific natural resources (including the topographical conditions and the movement of rainfall for hydro-electric energy).
3) The distribution of the population and of industry, which account for the size and type of demand.
These three factors all played their part in shaping the outlines of the Italian electrical industry during the early years of its history.

In Italy, the first commercial electricity plant was opened in 1883, at S. Radegonda, in Milan, operating with an Edison thermal dynamo. As was the case in other countries, it was initially used mainly for lighting. However, contrary to the experience of other nations in which thermal generation predominated, Italy soon specialized in hydro-electric generation, thanks to the availability of water resources – above all from the Alpine Basin – and a parallel shortage of coal which the availability of a large supply of electrical energy promised to overcome[3]. The Italian electrical industry developed large plants at an early stage. Already, by 1898, a plant of 14880 kW, Vizzola, was in use in Northern Italy, and high-voltage transmission lines carried cheap hydro-electric energy to bulk consumers in northern Italy. The technology that was selected for this purpose was the three-phase ac current which, as we shall see below, posed significant problems for mechanical uses.

Ac currents were largely adopted for thermal uses of electrical energy, replacing coal in the production of cast iron and steel in electrical furnaces, despite the heavy losses during transformation. The extensive use of electricity was not due to its efficiency compared to coal, but because it was a cheap internal substitute for imported coal. In 1913 Italy already produced 2% of the world total of steel produced by electricity, against Germany's 0.5% and the United States' 0.9%. In 1930, it reached 12% against Germany's 0.9% and the United States' 1.5% (see Table 1). Electrical energy was used just as much in the industrial "filieres" associated with it: in the production of aluminium, nitrogen and fertilizers, for example[4]. Italian entrepeneurs introduced several innovative processes for the

future of economics, Technical Report n. 533, Institute for Mathematical Studies in the Social Sciences, Stanford, California, August 1988.

[3] At the end of the 19th century there was a strong debate on the opportunities opened to Italian industry by the large availability of cheap electric energy to substitute for coal; cf., for example, F. S. Nitti, La conquista della forza, Bari 1905.

[4] On the interdependences between the electrical sector and other sectors of Italian industry cf. R. Giannetti, La conquista della forza: risorse, tecnologia ed economia nella industria elettrica italiana (1883–1940), Milan 1984, p. 139.

Table 1: Production of Electrical steel in selected countries (1913–1937) %

Anni	Paesi			Mondo
	Italia	Germania	Stati Uniti	
1913	2	0,5	0,09	0,2
1925	12,5	2,5	1,4	1,4
1926	13,9	2	1,4	1,4
1927	12,5	1	1,5	1,4
1928	12,2	0,9	1,6	1,6
1929	11,8	0,9	1,7	1,6
1930	12	0,9	1,5	1,7
1931	12,8	1,1	1,6	1,8
1932	19,3	1,4	1,7	2,4
1933	21,4	1,7	1,8	2,6
1934	21,1	2,5	1,4	2,6
1935	25,2	2	1,6	2,9
1936	29,2	2,5	1,6	2,8
1937	28	3,3	1,7	3

Source: G. Vignuzzi, Le applicazioni alla siderurgia, in: Supplement to n. 4 of L'Elettrotecnica, 26, 1939, p. 198.

production of steel in electric furnaces as Ernesto Stassano did at the turn of the century (1898), or to produce nitrogen as Giacomo Fauser did in the 1920s.

The technology (ac threephase) selected by the Italian electrical industry, along with other factors, hindered the diffusion of electric motors in industry, especially in the machine tools field. The threephase ac current at high voltage was the best way to transmit electricity over long distances, but had an important disadvantage for final users who used electricity to power electric motors. It allowed only one (later up to four) speeds which were not adequate to drive complex machinery such as machine tools. In this case problems arose from the machines which were used to drive a variety of tools. The figures of the Census of 1911, 1927 and 1936–38 confirm these observations. The distribution of electric motors in Italian industry shows a percentage for the mechanical sector closer to that of the textiles and metallurgical industries than in other nations (see Table 2). The figures referring to the kW per worker in 1927 and in 1936–38 are also significant (see Table 3). For both dates, the energy-intensive sectors (metallurgy, chemistry) had figures higher than the mechanical rather than the engineering industry as one might expect.

Table 2: Electric motors in Italian industry at Census (1911–1927–1936/37) kW

Branch	1911	%	1927	%	1936/37	%
Mining	8103	1,489035	68913	2,406264	174356	3,275529
Timber	24527	4,507165	109960	3,83952	190047	3,570307
Food	50935	9,359989	331727	11,58305	651141	12,23262
Paper	20077	3,689418	73978	2,583121	188356	3,538539
Iron and metalworking	39855	7,32389	407983	14,24571	884030	16,60778
Not metalliferous	7320	1,345148	167102	5,834771	359009	6,744501
Mechanical equipment	93207	17,12804	3521151	12,2962	1177033	22,11226
Textiles	98010	18,01065	585688	20,45002	773982	14,54037
Chemicals	32294	5,934455	149254	5,211565	508783	9,558222
Buildings	33284	6,116381	57752	2,016551	127006	2,385991
Others	134655	24,74466	557485	19,46594	289245	5,433884
Total	544178	100	2863900	100	5322988	100

Sources: Census, 1911, 1927, 1936/37.

Table 3: Electric kW per worker at Census of 1927 and 1936/37

Branch	1927	1936/37
Mining	1,6	1,28
Timber	0,38	0,67
Food	0,96	1,13
Paper	1,61	3,55
Iron and metalworking	3,33	12,07
Not metalliferous	0,97	1,75
Mechanical equipment	0,73	1,42
Textiles	0,51	0,64
Chemicals	1,5	4,49
Buildings	0,17	0,1

Source: Renato Giannetti, La conquista della forza, p. 164.

2. *Hydro-electric Energy and Self-production: The Early Days of Power Stations in Italy*

The trend towards power stations driven by hydro-electricity was an early and unique development. From two surveys taken by the Ministero dell' Agricoltura, Industria e Commercio (Ministry of Agriculture, Industry and Commerce), it is possible to see that hydro-electricity already made up 69% of installed power by 1908. Plants were restricted to exploiting low head streams, which provided loads limited to a few hours daily, and depended on the technical possibilities of the combination of generator, turbine and fixed paddle.

By the end of the century several large water power stations were operational; Paderno (1896) with an output of 9,555 kW, Caffaro (1898) 7,717

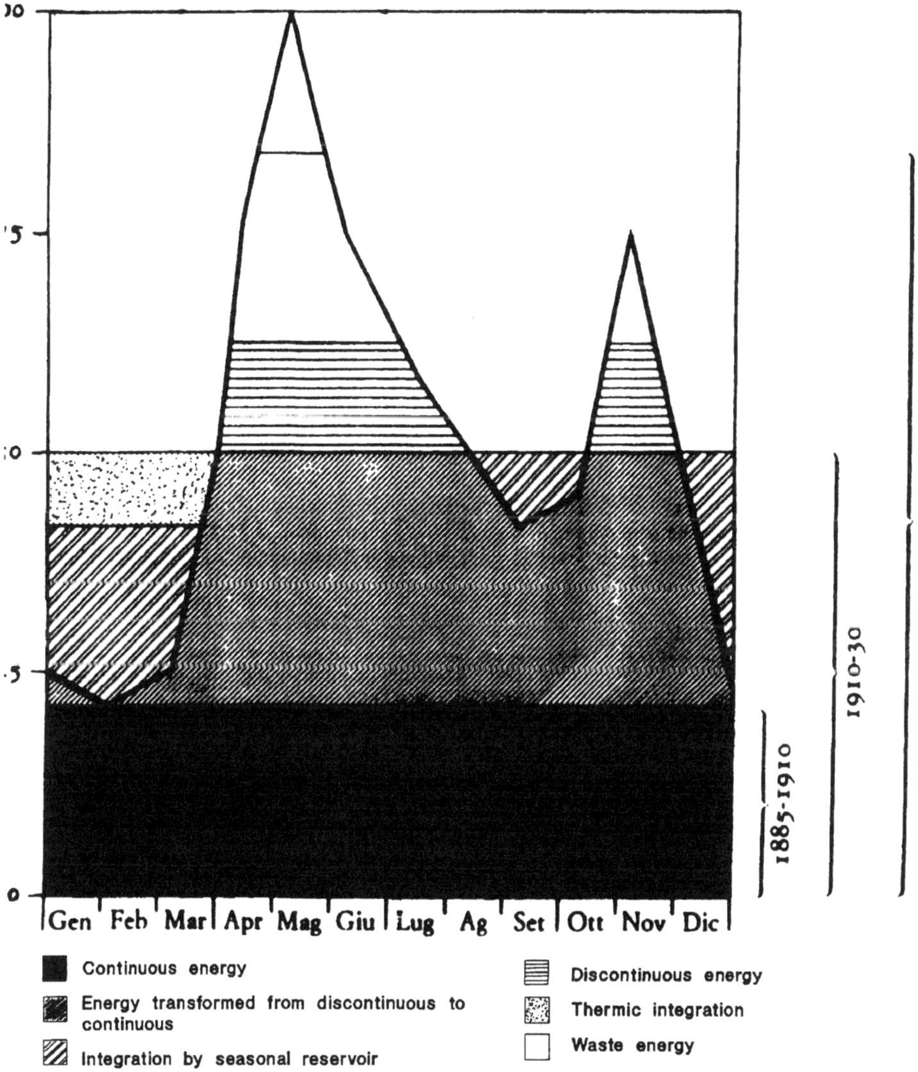

Figure 1: Monthly Capacity of Alpin basins and energy supplied

Source: R. Giannetti, I sistemi elettrici italiani, in: B. Bezza (ed.): Energia e Sviluppo. L'industria elettrica italiana e la societa Edison, Turin 1986, p. 290.

kW, and Vizzola (1898) 14,700 kW. Overall, the average output of each installation grew from 2,000 kW to 6,843 kW, between 1883 and 1910, even though the most common class were stations generating between 500 and 1,000 kW[5].

The relative ease of installation, and the availability of industrially exploitable falls, encouraged the formation of a system in which small and self-producers – to be found in the textile and paper production industries in particular – took advantage of the ready availability of resources. The Ministry of Agriculture, Industry and Commerce survey quoted above, shows that during the period 1883–1910, the percentage of self-producers with plants above 200 kw (300 Hp ≈ 225 kw) was still around 50% of the total power installed. The profits of small scale production were due to the availability of free resources, which compensated for the greater cost of installation demanded by hydro-electric stations as opposed to thermal ones.

By adding the technical difficulties of transmission, transformation and reduction of current to the factors discussed above, a full picture of the constraints imposed on the establishment of an integrated network by the early trend towards hydro-electric generation can be established. The greatest of these constraints was posed by the proportioning of the hydraulic diagram to the shallow level of capacity. This imposed certain technical ties which were incompatible with a "network view" of the production of electricity that emerged in the 1920s. Plants with extremely low rates of use of effective capacity were built, though these kept the falls in check all the same; canalization and building works took place which afterwards weighed heavily on the cost of reproportioning according to the new view of an integrated network.

The system of legal regulation during this first phase of development corresponded with the contemporary structures of enterprises, that is, it was fragmentary and incomplete. Legislation was restricted to laying down regulations relating to hydro-electric works, starting with law of 20 March 1865, n° 2248, which controlled the diversion of public water for the first time. It was revised with the law of 10 August 1884, and then brought to completion through several changes (laws: 30 March 1893, n° 173; 7 July 1902, n° 304), the final one being that of 29 July 1904, n° 923. The concern voiced constantly in these laws was the need to integrate the work necessary for the production of electricity with the efforts which had been made to protect the rivers and basins. These had been enforced

[5] For this data cf. ibid., Appendix 1.

on behalf of the agricultural owners who, up to then, had enjoyed rights of ownership and use of the water. Legislation was therefore directed mainly towards establishing forms of coordination between diversions and navigable canals, and between electrical and agricultural uses. This latter was particularly important since the difficulties and costs of long distance transmission of energy were forcing the diversion of water, where this was often detrimental to pre-existing agricultural interests. Particular difficulties were posed by the procedures relating to the issue of permission, which the law of 10 August 1884 delegated to the Prefects. Thus it was that the same river or basin often fell under different administrative jurisdiction, and that, as a result of this, more than one concession on the same resource could be given.

For the time being, however, concessions were bought up and plants were built under the influence of the electrotechnical manufacturers who competed to enlarge the market for their equipment. The development of consumption and the rationalization of transmission would only later be taken into consideration.

3. Early Regional Systems of Power (1912–1922)

The plants built during this period (1912–1922), were suitable for the flows of water available for seven or eight months of the year. The periods of shortage were moderated by the use of weekly and/or seasonal dams, and with thermal power stations, enabling some hydro-electricity to became continuous. During this phase, relatively few power stations were built, but the most widespread output capacity increased from 5,000 to 10,000 kW[6]. Attempts were made to gather the concessionary agents in each area around several of the larger power stations which were capable of covering the base-load, and backing these up with smaller power stations during peak-periods.

It was as a result of these new technical possibilities, and of the industrial strategies that were consistent with them, that the lower amounts of construction typical of this period came about. In the ten years between 1912 and 1922 the average total power grew only 9% per year, in contrast with the preceding ten years, when the average rate of growth was 16%

[6] Cf. R. Giannetti, Tecnologia, scelte d'impresa ed intervento pubblico: l'industria elettrica italiana dalle origini al 1921, in: Passato e Presente, 2, 1982, p. 61.

per year (see Table 4 and Figure 2). It was not, in fact, a case of making progress as a result of building new capacity, as much as moving ahead through nationalisation, and by providing the kind of business policies needed by the respective networks[7]. These strategies gave rise to relatively large regional systems, especially in northern Italy, that placed Italy fourth on a world scale, and first in Europe[8] among those countries which were equipped with plants transmitting more than 70 kV.

Varying interests were involved in this operation, giving rise to several struggles, described elsewhere as "parallel wars"[9]. In these "wars" there

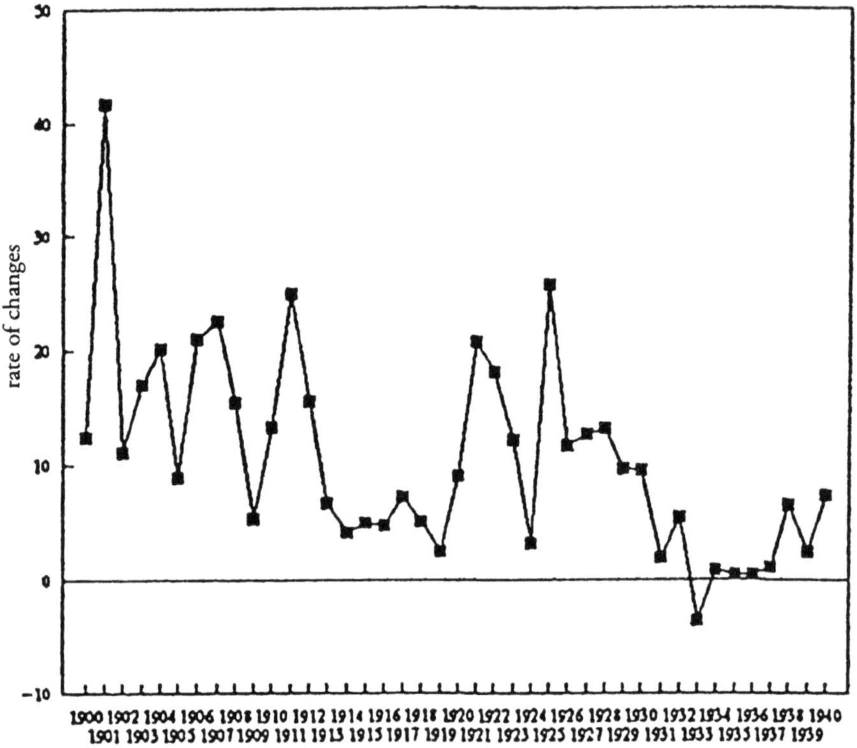

Figure 2: Rate of change of total power in Italian Electrical Power

[7] Regarding the progress made in hydraulic construction cf. L. Rushmore, Hydroelectric Power Stations, New York 1917.

[8] Cf. S. Haar, Supplement to Electrical World, 4, 1914, republished by Bureau of the Census, Central Electric Light and Power Stations, 1915, p. 132.

[9] G. Mori, Le guerre parallele, in: Studi Storici, 14, 1973, p. 292, now in: G. Mori (ed.), Il capitalismo industriale in Italia, Rome 1977, p. 141.

Table 4: Total power of Italian Electrical Utilities (1899–1940

Anni	Total power	Rate of Change %	
1899	113		
1900	127	12,38938	
1901	180	41.73228	
1902	200	11,11111	
1903	234	17	
1904	281	20,08547	
1905	306	8,896797	
1906	370	20,91503	
1907	453	22.43243	
1908	523	15,45254	
1909	551	5,353728	
1910	624	13,24884	
1911	779	24,83974	
1912	900	15,53273	15,89711
1913	960	6,666687	
1914	1000	4,166867	
1915	1050	5	
1916	1100	4,761905	
1917	1180	7,272727	
1918	1240	5,084746	
1919	1270	2,419355	
1920	1385	9,055118	
1921	1672	20,72202	
1922	1973	18,00239	8,971303
1923	2214	12,2149	
1924	2284	3,161698	
1925	2870	25,65874	
1926	3206	11,70732	
1927	3613	12,69495	
1928	4088	13,14697	
1929	4485	9,71135	
1930	4912	9,520624	
1931	5001	1,811889	11,76288
1932	5276	5,4989	
1933	5083	−3,65807	
1934	5123	0,786937	
1935	5145	0,429436	
1936	5166	0,408163	
1937	5219	1,025939	
1938	5581	6,552979	
1939	5700	2,49955	
1940	6119	7,350877	2,270066

Source: Renato Giannetti, La conquista della forza, Milano 1985, p. 255.

were those who wanted a complete integration of electricity production and the development of the networks – the so-called "electrobankers" were among these – and there were those who wanted independence from the commercial electricity enterprises. This group included chemical producers and the electrometallurgists, who feared tariff increase resulting from investment in coordination. Finally, there were those on the edge of the larger industrial conflicts, who did not want to lose the chance of speculating.

The war proved a great stimulus to all this, and the money from public orders, together with fiscal exemption for the reinvested extra profits, provided the financial resources to support it. As the intricate financial affairs of the period show, the overall result was not an integrated network of large plants, but a system which was fragmentary on the industrial level, and unstable from a financial point of view. The state played absolutely no autonomous role in coordinating this, but was instead seen as ground to be conquered by the conflicting lobbies. As a result, a long and complicated process led to the producers coming together into nine basic groups; Società Idroelettrica Piemontese, Edison, Adamello, Societaà Adriatica di Elettricità, Unione Esercizi Elettrici, Tridentina, Ligure Toscana e S. E. L T. Valdarno, and Società Meridionale di Elettricità (see Figure 3). It also led to the consolidation of a substantial group of self-producers, above all in the chemical and electrometallurgical sectors (Falck, Montecatini, Caffaro, etc.).

4. *The Construction of a National Network in the 1920s: A lost chance*

In the immediate post-war period, a new phase of electrification began. Its origins were to be found in the final stages of the previous era, based as it had been on lighting and limited thermal uses. The new era heralded the rationalization of production through the construction of transmission networks.

In Italy, the fact that the big consumer centres for hydro-electric power were generally far apart, made the idea of moving towards integration and interconnection very attractive to commercial producers. However, the Italian system was still dominated by those particular factors which had enabled the initial expansion to take place. The different periodicity of transmissions and the great number of lines made it hard to establish linkages between systems. Coordination and exchange between local systems required many modifications throughout the system, including both lines and equipment.

Figure 3: Regional distribution of power producing groups

The main problem was how to connect different local systems. There were essentially two solutions. The first was to proceed to a complete unification of frequency and tension, the second was to install groups of converters at key points in the network, the gateways, without modifying the different systems. To adopt the "jargon of the theory of standard"[10], the problem was the setting of standards for interface compatibility. The

[10] On this cf., for example, P. David, Some New Standards for the Economics of Standardization in the Information Age, in: P. Dasgupta/P. L. Stoneman (eds.), Economic Policy and Technological Performance, London 1987, ch. 7.

crucial point was the frequency of the current. In Italy there were at least two standards, 42 and 50 cycles. The problem was to evaluate the consequences of a complete change of frequency in a certain region – from 42 to 50 cycles and vice versa – on existing transmission equipment (lines, wires) and on users' equipment (motors, pumps, electrofans, etc.). Alternatively, the effects of the installation of converters to exchange energy between different local systems needed to be explored.

Both alternatives had important drawbacks. In the case of a change of frequency, there were two different situations: one, with the same tension and one with a proportional increase or decrease of tension. In the first case, the main effects of an increase in frequency in the transmission equipment were a deterioration in the regulation of the line and a slight increase in waste of dielectrics and of conductors. As for the users' equipment, the higher frequency increased the number of revolution in electric motors so that the speed of machine tools rose. This could require the modification of the pulley or, in machines directly coupled to motors, the substitution of the motor itself. Problems also arose in generation plants. For example, the generators decreased their efficiency somewhat but maintained their power. On the other hand, the thermal-engines increased their power efficiency, but, generally, could not cope with the corresponding increase in speed. Important drawbacks could arise from lowering the frequency from 50 to 42 periods. In the field of generation equipment there was a small decrease of power, but this could increase the temperature, damaging the plant. The effects on lines and wires were less important, they implied a general decrease in wastage, but also a decrease in the power-factor.

In the case of a simultaneous increase in frequency and tension, generation plants, electrical equipment, transmission lines and wires had almost the same problems as in the case of constant tension. More significant problems could arise by diminishing frequency and tension. For example, wastage grew, increasing operating costs.

In the case of the adoption of frequency converters, two systems have to be distinguished: a converter together with a motor and a synchronous generator, and a converter with an induction motor and a synchronous generator. Both had advantages and disadvantages. The first type of converter worked in both directions without requiring variations in the ratio of the frequencies of the systems connected, allowed easy control of power-factor, and could be protected against overload. On the other hand, this last feature made it easy to interupt the interconnection, and compelled more frequent synchronization of its components. The second type had the advantage of avoiding a surge, but limited the connection to two local networks at a time.

The Italian Electrical Engineering Association (Associazione Elettrotecnica Italiana), suggested a very cautious program of progressive advancement toward the standardization of frequency, proposing a two stage project. In the first phase, the electrical enterprises had to unify their respective zones of influence, with special attention to new plants, which had to be built in accordance with the unified frequency. Only later would they proceed to a wider integration at a national level. This program excluded a more complete integration, where the plants were projected on the basis of the needs of a national load prefering, the belief that different systems had to be connected only for auxiliary purposes or in emergency situations. This phase of transition was required because of the difficulty for most Italian local systems of proceeding to faster unification. Most of the major plants produced the 42 cycles frequency, and could not be converted to 50 periods, but could be easily upgraded to 46 without major or expensive improvements to their equipment. Moreover, such a small reduction did not damage the efficiency of the 50 cycle systems.

Besides these features of a technical nature, there were corresponding economic and business problems. Up to that time, efforts to coordinate had principally taken the form of the division of areas of influence and tarif cartels, to the detrimental effect on the efficient coordination of the lines.

As a result it was difficult to arrange for efficient organisation of operations. For example, the delivery of energy in the order of a few thousand kW were possible only with voltages of 10 to 60 kV. For higher voltages of 200–900 kV, the receiving stations were profitable only at a scale of 50,000 kW, far from consumption levels at any station in immediate post-war Italy[11].

It was the relative lack of demand which rendered the construction of large plants rather unappealing, given that to be economical they would have had to feed many centres of consumption. However, in the contemporary Italian situation, these centres often had autonomous generators of small capacity at their disposal, and they restricted use by linking to larger plants only in emergencies through ad hoc lines. The resulting de-

[11] Cf. G. Sartori, A proposito di una rete unica di trasporto e distribuzione dell'energia, in: L'Elettrotecnica 5, 1918; Ignis, A proposito di una rete unica di trasporto e distribuzione della energia e della questione del sistema, in: ibid.,p. 356.

mand was thus extremely limited and variable, and could not cover the expenses necessary for integration into a network. For this reason, the Italian Electrotechnical Association maintained throughout the 1920s that the existing system of exchange, although giving rise to substantial losses of energy owing to the presence of convertor groups, was to be preferred from an economic point of view. This was because the expenditure required for the replacement of machinery, both in power stations and users' homes, which would be necessary in the event of the unification of frequencies, was considered unsustainable.

State intervention, in this phase of Italian electrification, was very prominent, but its effects were only modest. The most significant piece of legislation was the Bill of 12 February 1919, signed by the Minister of Public Works, Bonomi. It provided uniform measures regarding public participation in firms, and tariffs and grants for the construction of dams.

It was here that the major contradictions between what the law prescribed and what was actually put into practice became apparent. The Bill was intended to promote the construction of dams in order to regulate flows, allowing greater flexibility in the use of hydro-electric energy. The subsidies were distributed on the basis of the capacity of the dams, and on the financial deficit shown by the concessionary companies at the end of the work. In relation to the first point there was a tendency to build the dams larger than was actually required according to accurate industrial estimates, for the simple reason that the cost grew more slowly than the capacity, so that the larger the dam, the greater was the profit for the concessionary firm. In relation to the second point, the costs indicated by the firms grew at such a rate that, in 1922, the Principal Council of Water Commission (Consiglio Superiore dei Lavori Pubblici) proposed immediate and radical alterations, aimed at changing the criteria for benefits from ex post to the ex ante value of costs. The Council itself had to estimate and check them[12].

The second measure of support for the electrical industry was proposed by Minister Corbino, and granted subsidies for both production and, more importantly, transmission plants. These permitted an efficient system of exchange to be established. This Bill also altered significantly in practice; subsidies meant for transmission and distribution were awarded to power lines of both medium and large size without strict control, thereby dissi-

[12] Regarding the result of ddl. Bonomi, cf. Ministero dei Lavori Pubblici, Consiglio Superiore della Acque Pubbliche, Statistical report on the enforcement of the Bill of 20.11.1916 n°1664 and alternations, Rome, 1921–23, 2 vols.

pating the resources amongst all the concessionary firms, without specific reference to long-distance transmission lines.

The geographical shape of the Italian peninsula required a longitudinal transport line to be constructed from north to south in order for a national electricity network to be efficient. As has already been demonstrated, there was considerable technical and economic uncertainty over the question of proceeding to a greater level of integration. The principle obstacle was the enormous amount of financial resources needed. The line would have to be able to carry the maximum power load required by the network, but, at that point in time, the load-factor was at only 31% not nearly enough to ensure a sufficient return to justify financing the investment. In order to obviate these difficulties the promoters of the a "backbone" proposed several ideas for public support of the project.

According to them, the primary distribution network would be built by the state and run, under a concessionary structure, by private electric companies. As to finance, they proposed a special body, the Electric Bank, to look after this, funded by the imposition of a charge levied on the basis of aggregate consumption of energy[13].

This project also provided for the State to lay down binding procedures regarding the use of basins, the frequency of currents and the kind of interconnection between the different systems, by means of a regulatory plan. The bank's task would be that of financing those projects which corresponded to the norms laid out in this plan. This proposal was an attempt to substitute the uncontrolled action of the various banks with clearer criteria for the assignment of credit, for state intervention in the control of the flow of funds necessary for financing investments, and to insurance the stocks which were to be the guarantee.

The project met with much opposition, above all from Edison. This group opposed state intervention, prefering instead the free enterprise system, and maintaining that contemporary conditions were already restrictive as a result of existing public constraints[14]. This kind of opposition was not new, nor was it particularly original. In reality, Edison's hostility was intended to stop a new wave of construction placing old plants at a disadvantage with respect to costs and charges. This is obvious from the fact that soon afterwards Edison joined with all the other electricity com-

[13] Cf. P. Bignami, Per un grande Istituto di Credito per imprese elettriche, in: L'Elettrotecnica 6, 1919, p. 511.

[14] Cf. D. Civita, Per un grande Istituto di Credito per imprese elettriche, in: L'Elettrotecnica 6, 1919, p. 562.

panies to request that the regulations of the law of 20 August 1921 (regarding public contributions to the hydro-electric plants to be built in the south) be extended to the whole nation[15].

An important part in the failure to construct a national network was played by the events which upset the international market for electrotechnical plants during the years 1920–1925[16]. It was in precisely this period that national and continental areas lost their distinguishing competitive features. This was due to massive international integration, in particular, as a result of the crisis among German firms and of the penetration of the big American companies, General Electric and Westinghouse.

In Italy, the many-sided conflict over oligopoly which was taking place on an international scale contributed to the failure of the construction of the networks, especially through the reorganisation of the Swiss company Brown-Boveri[17]. It was the most active in this field, working through Sip (Società Idroelettrica Piemontese) and also through the Conti group. The sudden change in the balance between manufacturers of electrical machinery opened the Italian market up to many foreign producers, for example the Swedish newcomer, Allmanna Svenska. It also opened it up to many small producers who exploited this or that patent, in contrast with the project for coordination. This required extensive standardization of equipment to reach the dynamic and static economies which existed during the actual phase of the growth of production of electricity, which was occurring simultaneously in small dynamic economies such as Sweden and Switzerland.

If it is true to say that the conflict between the large oligopolistic electrotechnical firms contributed to preventing the construction of a national electric system up to 1925, it is just as true to say that the final blow to this plan came in the form of the monetary measures adopted between 1925 and 1927.

Post war inflation had in fact helped firms with an older stock of capital, since they had realized substantial advantages in capital accounts

---

[15] Cf. D. Civita, Per il finanziamento delle imprese elettriche, in: L'Elettrotecnica 9, 1922, p. 481.

[16] Regarding this aspect, cf. R. Giannetti, The Conflict among Electrical Manufactures in the Twenties and the Italian Case, EUI Colloquium papers, Florence, 17–19 October, 1984.

[17] For this aspects cf. BEAMA, Combines and Trusts in Electrical Industry, the position in Europe in 1927, London 1927; E. Hess, Elektropolitik und Weltverstromung, Amsterdam 1931; A. Gebhard, Die Expansion der Amerikanischen Elektrokonzerne in Europa, Heidelberg 1932; Board of Trade, International Cartels and Internal Cartels, 2 vol., Washington 1944.

in relation to the part of this which was financed with debts. In contrast those firms which were building under the new conditions experienced a more costly global market. The latter were seeking the liquidation of older capital in order to leave space for the financing of new enterprises[18].

Here too, Edison and Sip were in conflict. Edison's main construction effort had taken place before the War, and the adoption of a revaluation strategy of the plants would help to revive profitability, which was impaired by relative industrial backwardness. This was not true for Sip, which had, and was continuing to install, modern machinery. In its case, registering the plants at real prices meant making fiscally disproportionate depreciation, with negative effects on the policy of dividends. The best strategy would be that of expanding sales, which allowed them to input to the current budget the new plants' expenses, with the underestimation of the book keeping figures of these.

The condition of stabilization helped Edison considerably. In 1927, it had gained so much ground over the other electric firms that its deputy administrator, Giacinto Motta, took up office as president of the sector's Association, Unfiel (Unione Nazionale Fascista Imprese Elettriche)[19]. Simultaneous with the process of stabilization was the acceptance by public authorities of the idea of a "national network" in 1927[20].

On this occasion the Principal Council of Public Works rejected the project of building a line capable of transporting energy throughout the country. It maintained instead that the energy from the various groups would be exportable only to the extent by which it exceeded the requirements of the district concerned. The region was felt to be the appropriate unit for dealing with all needs: those of the electricity companies, those of agricultural communities, and social ones – all appeared to be under threat from the transfer of resources permitted by a national system of transmission.

---

[18] Exemplary, in this sense, is the debate of 1927 between G. Motta and L. Federici. Cf. L. Federici, Gli investimenti industriali in regime di svalutazione, in: Gioranale degli economisti 6, 1927; G. Motta, Gli investimenti industriali in regime di svalutazione, in: ibid., afterwards republished again in L'Energia Elettrica 24, 1927, p. 1095.

[19] The First coordination between Italian electric firms came about in 1920 with the name Associazione Esercenti Imprese Elettriche (AEIE), with its own magazine, L'Impresa elettrica. In 1923, it was Motta who formed the rival association Associazione Nazionale Industrie Elettriche (ANIEL) which had its own publication, L'Energia Elettrica. In 1925, the two associations joined together forming Unione Nazionale Industrie Elettriche (UNIEL), which became UNFIEL in 1927.

[20] Consiglio Superiore dei Lavori Pubblici, Trasmissione e scambio di energia elettrica. Parere emesso dall'Assenblea generale del consiglio Superiore dei Lavori Pubblici nella adunanza del 28 dicembre 1927, in: Annali dei Lavori Pubblici 1, 1928, p. 97.

5. The Network in the Thirties

It was not until after the failure of the big general bank, and the rearrangement of Italian electrical groups that the construction of a major line linking the Alpine hydro-electric system with those of the centre and the south again became a topic for discussion.

The context had, however, changed completely. The question of the network had earlier been raised in the light of a deficit of energy for northern industry, and it was for precisely this reason that the latter intended to link up with southern resources. In 1934, however, the defining factors had been reversed. After the depression it was the large northern groups (Edison and SADE) which sought to build the "backbone" to transport their excess energy towards the south, and not vice-versa, as had been projected during the twenties. But on this occasion too, the enemies of the "backbone" emerged victorious, and the question was not addressed again until after the Second World War.

The German Long Distance Telephone Network as a Large Technical System, 1919–1939, and its Spin-offs for the Integration of Europe

HARM G. SCHRÖTER

1. *Introduction*

From the late 1920s onwards Germany possessed the most advanced network for interstate telephone communications in Europe. Before the First World War only a few reliable international telephone lines existed[1]. This was transformed by construction of a large network of underground cable lines in which Germany took the lead, and which grew very quickly. In 1920 less than 300 km of cables were in use, this had risen to 10,000 km in 1930, and had reached nearly 20,000 km, which represented 6 million km of single lines, in 1938[2]. This growth is astonishing given that demand remained at the same level from the early 1920s onwards. In 1922, when demand was at its highest, 315 million long distance calls were made in Germany, a number which was not surpassed until as late as 1938, with 353 million calls[3]. What caused the Deutsche Reichspost, which operated the telephone system in Germany, to invest so heavily during the whole period despite stagnating demand?

Drawing on Thomas Hughes' approach (see below) to Large Technical Systems (LTS)[4] this paper will examine why and how this network was

[1] Cf. N. H. Wasserman, From Invention to Innovation: Long-Distance Telephone Transmission at the Turn of the Century, Baltimore 1985.

[2] Cf. Deutsche Reichspost: Verwaltungsbericht über das Rechnungsjahr 1938, p. 65.

[3] Cf. K. Sautter, Geschichte der deutschen Post, Teil 3, Frankfurt/M 1951, supplement, table "1920 bis 1943".

[4] Cf. T. P. Hughes, Networks of Power: Electrification in Western Society, 1880–1930, Baltimore 1983; T. P. Hughes, The Evolution of Large Technological Systems, in: W. Bijker/ T. P. Hughes/T. Pinch (eds.), The Social Construction of Technological Systems, Cambridge 1987, pp. 51–82.

constructed, and to what extent it helped – willingly or not – with the integration of Europe. This contribution has three aims. Firstly it describes and discusses the extension of the German cable network. The second aim is methodological, transfering Hughes' model of LTS to a subsystem of a LTS. Thirdly it contributes to economic history, by investigating the extent to which technical supply-side opportunity was taken up by the economy.

A spin-off of adopting the LTS approach is the implications which emerge for the integration of Europe. The question should be seen in the light of current policy. Bureaucrats and politicians working for the integration of Europe pursue their aim in several ways. One of them is to use technical means to narrow the room for political manoeuvre to such an extent that in the end political decisions are reduced to "how" rather than "yes" or "no". Given this background it is interesting to see the extent to which the technical integration of Europe through the development of the telephone network in the interwar period stood for a framework of technical possibilities, and influenced political decisions through its existence.

By integration we mean the development of a whole by bringing together separate elements. In this case this means the construction of a single European telephone network. The individual national networks, initially separate elements, were connected to each other, until in 1938 the subjects of all European countries were able to contact one another by telephone[5]. But technical possibilities are only one side of the coin, while their use represents the other. Therefore, after outlining the spread of the technical system, I will discuss how it was used, since integration is not just about opportunities but also concerns reality.

In order to describe and interpret the growth and extension of technical communication systems authors have approached the topic from several directions. In his discussion of the international telegraph system up to 1914, Jorma Ahvenainen saw "success" as based on economic or political factors, or on a combination of both – without offering much further interpretation[6]. Horst A. Wessel stressed technical development in his

---

[5] In 1938 Germany had telephone connections directly or indirectly with all European states except Albania. This offered the possibility for all other states to contact each other through the German network. This opportunity was not always used for political or economic reasons. But the direct connection e. g. between Sweden and Finland from Norrtälje via Mariehamn to Turku had broken down, it would have been technically possible to call via Germany, Lithuania, Latvia and Estonia.

[6] "The only builders of intercontinental telegraph lines to really succeed were Britain, Denmark, Russia and, for the American continent, the United States. Britain's sucess was

book on German telecommunication before World War I[7]. Hartmut Petzold emphasized "German-French Rivalry and Competition" as a driving force for the erection of the European telephone network[8]. Only Frank Thomas has applied the concept of the LTS to the growth of the German telephone system[9]. This approach is adopted here because it offers the most fruitful one within which to discuss our questions[10].

There is no doubt that the German telephone system represented a LTS. The cable network was part of the telephone system. It can, therefore, be described as a sub-system, but was itself large in terms of invested capital, number of employers, and sheer size[11]. It was technical, since information was carried by means of electrical waves, and it was a system because various apparatus of different kinds (not only several cables) had to be connected in a particular way in order to make it work. It was clearly defined in relation to other parts of the telephone system, and kept an

---

both political and economic, Denmark's was economic and Russia's political." J. Ahvenainen, Telegraphs, Trade and Policy. The Role of the International Telegraphs in the Years 1870–1914, in: L. R. Fischer (ed.), Shipping and Trade, 1750–1850, Pontefact 1990, pp. 505–18; idem, The Far Eastern Telegraphs, The History of Telegraphic Communications between the Far East, Europe and America before the First World War, Helsinki 1981.

[7] Cf. H. A. Wessel, Die Entwicklung des elektrischen Nachrichtenwesens in Deutschland und die rheinische Industrie. Von den Anfängen bis zum Ausbruch des Ersten Weltkrieges, Wiesbaden 1983.

[8] H. Petzold, Deutsch-französische Rivalität und Zusammenarbeit bei der Errichtung des europäischen Telefonnetzes nach dem Ersten Weltkrieg, in: Y. Cohen/K. Manfrass (eds.), Frankreich und Deutschland. Forschung, Technologie und industrielle Entwicklung im 19. und 20. Jahrhundert, Munich 1990, pp. 263–80.

[9] Cf. F. Thomas, The Politics of Growth: The German Telephone System, in: R. Mayntz/ T. Hughes (eds.), The Development of Large Technical Systems, Frankfurt 1988, pp. 179–213.

[10] Cf. H. G. Schröter, Innovationsverhalten und technologische Entwicklung: Zu den Gründen für die fernmeldetechnische Vorreiterrolle Deutschlands 1920–1939, ms.; Other authors, too, have evaluated telecomunication using similar approaches. E.g. Bar and Borrus have looked into the three "layers" of telecommunication, which correspond with Hughes' approach: Their "physical network" corresponds with Hughes' reverse salients, "mangement" with momentum and "demand" with load factor (F. Bar/M. Borrus, From Public Access to Private Connections: Network Strategies and Competitive Advantage in US Telecommunications, in: Berkely Round Table on the International Economy Organization for Economic Cooperation and Development (OECD), Information Networks and Competitive Advantage, Berkeley 1989, Vol. 2, pp. 5–61.

[11] In 1939 119,053 km of cables representing 25,600,107 km of single lines were counted (Austria still excluded, but inner-city connections included). Cf. Deutsche Reichspost, Verwaltungsbericht über das Rechnungsjahr 1938, p. 65. In 1930, when 57% of all local exchanges had already been automated, 22,613 people were employed in the exchanges alone. Cf. Deutsche Reichspost, Geschäftsbericht über das Rechnungsjahr 1929, p. 62.

organisation of its own[12]. Contemporaries looked at it as a separate system. There was even a periodical, "Das Fernkabel", a name later changed into " Europäischer Fernsprechdienst" (EFD), monitoring the growth of the cable network not only in Germany but in the whole of Europe. In what follows Hughes' model of LTS is applied not to the super-system of telephony as a whole, but to the cable network element of it. This case study will be used to demonstrate how this model applies not just to the LTS as a whole, but to defined large parts of it as well.

Hughes' model describes three phases in which a system is built up, and identifies three "structural features"[13], which are understood to act as propelling forces. Here the phases and features of the model are briefly outlined, followed by their application to the empirical development of the German cable network.

Phase one is an innovation, which itself includes the three steps of invention, development and spread, which brings the system into use[14]. The second phase represents the transfer of the system to a different environment, which may change the technological style. During the third phase the system passes from growth through competition to consolidation. As the system becomes rationalized, more efficient and more capital intensive, the skills required to run it change, too, from the engineer-entrepreneur to the manager- and the financier-entrepreneur.

The LTS is propelled by "reverse salients", "load factor", and "momentum". Reverse salients describes the structure of technical growth, which is by no means steady, because parts of the system are technically more advanced than others. This creates the incentive to develop the less advanced parts, which in their turn become the most advanced ones within the system. The weakest point is thus redefined, creating new demands and so on. Reverse salients focus on qualitative changes within the system, while the load factor primarily explains the quantitative expansion of the system. The load factor is the ratio of average use to maximum use of the system. When average and maximum use are narrowing, demand suggests the widening of the system, in our case the addition of new cable

12 The structure was quite complex: The cable was laid, maintained and owned by the private Deutsche Fernkabelgesellschaft. The Reichspost had to pay for the use of cables to the Deutsche Fernkabelgesellschaft. This enterprise was founded by the Reichspost, which was the major partner, together with German firms producing cables and telephones. But the amplifier stations, situated between parts of the cable as well as the exchanges were owned and managed by the Reichspost.

13 B. Joerges, Large technical systems: Concepts and issues, in: Mayntz/Hughes, p. 13.

14 Cf. F. Pfetsch, Innovationsforschung in historischer Perspektive. Ein Überblick, in: Technikgeschichte 45, 1978, pp. 188–233.

lines. In our example, it must be stressed that "maximum" is defined not just by a technical capacity of 100%, but the figure considered by actors to be "sufficient" demand (see below). Finally, momentum includes the external effects of the system. In addition, here it is understood to mean influences from outside exerted upon the system as well as effects imposed by it on the outside world, a picture which represents an addition to Hughes' model.

2. *Phases of Growth*

Cable networks were constructed from 1850 onwards for the telegraph system. Cables did not need to be invented anew for the telephone, but they did need to be restructured. The technical requirements for telephone cables were much higher than for the telegraph. Though a telegraph could easily utilize a telephone line, telephones could not be connected to the existing network of telegraph lines. A new and quite capital intensive network had to be erected, but the cable itself had already passed through the first phase of development mentioned above. This does not mean that there was no room for major advances. Before World War I the maximum length of a telephone line was limited. Even with the application of Pupin coils[15], which allowed considerable extension, it was limited to about 700–800 km[16]. This limit was tested by the construction of a line between Berlin and Milano (1,350 km) shortly before the war, into which the most advanced technology was built. Its service turned out to be so unsatisfactory, that it was used only in two separate parts.

It was the amplifier valve which precipitated the breakthrough. During the war American specialists built a considerable telephone network in France working with these amplifier valves. In Germany too this technique was used for military purposes. It enabled the German High Command to speak directly with its Turkish allies. This sheds light on the situation before and after the War. The expertise for technical progress and industrial capacity for communication systems was concentrated mainly in the United States and Germany. Because in the US capacity and demand was much greater in every respect, German industry tried to

[15] Though the Pupin coil could reduce the attenuation significantly, it had the disadvantage of reducing the transmitted frequency too. Heavily coiled circuits permitted a frequency band up to 1600 Hz only.

[16] Cf. H. Kröber/K. Heyer, Geschichte der Leitungs- und Übertragungstechnik, Teil 2, in: Archiv für deutsche Postgeschichte, 2, 1983, pp. 112–28, p. 112 f.

compete not so much on quantitative but with qualitative factors by concentrating on technically more advanced goods, which often were to be sold at a higher price. The amplifier valve was the prerequisite for a network of telephone cables for long distance communication. It represented the technical precondition for the integration of Europe through the directly spoken word, accessable to everybody and without the need for special training such as that required by telegraph operators.

The second phase of development, the transfer of the system to other environments, can be tackled briefly in this context. Cable networks for telephone use were constructed in the USA, and after Germany had taken the lead in the Old World, in all industrialized states of Europe.

The third phase, however, growth through competition, was extended over the whole of the interwar period in all European countries. It was not the telephone in competition with the telegraph, since this question was settled quite soon[17]. The cable was in competition with single lines on the surface. These lines, which were not buried, were much cheaper than a cable and many state and private companies began with them. However, they proved less reliable and caused technical difficulties. Electrical induction caused problems ranging from the life-threatening to mere jamming[18]. This last problem was most common when many lines were carried together. To avoid the jamming the lines had to be constructed in a very narrowly defined way, absolutely parallel and with crossovers at precise distances. This meant that in the cities the chimney-sweeps and sailors, who were hired during the early stages when the systems were built on rooftops, had to be replaced by more professional people, at a higher costs. Furthermore, in big cities the number of lines could not be extended any more, as all available space was already used up by thousands of wires. But the main factor which caused the switch from overhead transmission on masts to cable laid into the earth in the countryside, was reliability[19].

---

[17] It was not just access for everybody which favoured the telephone, but charges too. This is illustrated by the following example: In the 1930s the charge for a three minutes call from Berlin to Paris was RM 6.40. During this time a stenographer could write down 360–450 words. But the costs for a telegram with 400 words were RM 76.00, over than ten times more than for the telephone call. Cf. P. Craemer, Wesen und Wandlungen des elektrischen Weltnachrichtenverkehrs, in: Europäischer Fernsprech Dienst (EFD) 38, 1935, pp. 7–14, p. 8. Furthermore this example excludes the main advantage of the telephone: direct response.

[18] In cases where the telephone line was conducted parallel with the tube in cities.

[19] In some countries cables were put on masts as lines had been before. This solved the electro-technical problems, but not those caused by wind and weather. Because of the threat

In 1909 weather conditions caused the German telephone system to break down for a long period. Ice and snow had frozen on the surface lines, when a heavy storm came and made the masts collapse over hundreds of kilometres. The Reichspost decided, therefore, to build a cable network protected by the soil. In the following year the construction of the famous "Rheinlandkabel" from Berlin to Köln was started. But when the War stopped further progress in 1914, only the part up to Hannover was ready for use. Experience gained from this Rheinlandkabel and long distance communication during the war paved the way for a far reaching decision by the Reichspost. In 1921 a plan was agreed to construct within five years a whole network of underground cables for the use of the telephone system[20]. Construction was started immediately. In 1920 the total length of long distance underground cables[21] was 298 km[22], in 1925 it was 4,743 km and in 1930 it extended more than 10,000km [23]. In 1934 the cable network in Germany had grown to 13,000km, carrying 3.4 million km of telephone lines. These figures represent 32.5% of all European cables, and 34.0% of single lines in such cables[24].

The following maps show the plan of 1921 and the construction of the network in 1931. During this time the construction of cable lines took place according to demand[25]. Since Germany was not a centralized country, the cable network had a decentralized structure. In 1931 Frankfurt had the largest number of cable connections with seven. Berlin could count on six, as could Stuttgart and Nürnberg. Dresden and Leipzig followed with five connections each. It is clear from the European map for 1933, that individual national networks were shaped differently. In more centralized states cable lines, by focusing on the capital, formed a star.

---

of earthquakes some nations, as Japan does even today, carried parts of their cable networks on masts.

20 Petzold, p. 272.

21 The cable network inside the towns was already much bigger at that time.

22 Cf. Deutsche Reichspost, Geschäftsbericht über das Wirtschaftsjahr 1924, p. 72.

23 Figures in the reports are a little confusing, as up to 1929 length was given for the end of the year, but from 1930 onwards for the end of March. Deutsche Reichspost, Geschäftsberichte über das Wirtschaftsjahr (from 1927 onwards: "Rechnungsjahr", and from 1936 onwards "Verwaltungsberichte über das Rechnungsjahr").

24 Calculated from: Craemer, p. 11.

25 Thomas wrote of the "[...] demand-oriented and cost-sensitive expansion policy [...]" during the Weimar Republic era (p. 197), which is contradictory to his own statement that: "Because of the lack of competition the DRP (Deutsche Reichs Post) was able to concentrate its efforts on the technical improvement of the existing system and neglected the existing demand," (Thomas, Politics, p. 196 .f). Several contradictions are found in his contribution.

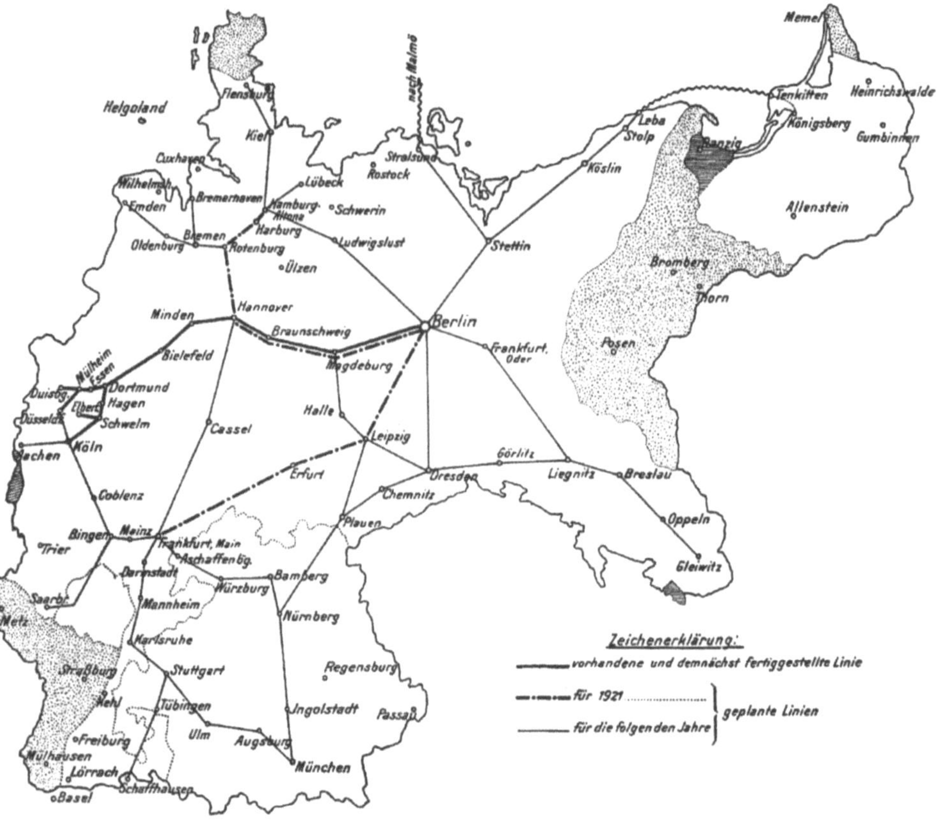

Figure 1: Existing and planned German long distance cable network, 1921
Quelle: EFD 54, 1940, p. 50

Brussels, for example, had six cable lines, while Liège the next biggest town, had only three. The same star-shaped concentration was to be seen in France (Paris:6, Lyon:3)[26]. When, in the second half of the 1930s the German cable network was prepared for war, there was no need to change the whole structure from a vulnerable star- into a more secure mesh[27]. The cable network, along with motorways and other large-scale construction projects, could be used for civil and military purpose alike.

---

[26] Planned cables, or those under construction, are not counted. Only cables, not lines inside cables were counted. Cf. Deutsche Reichspost, Geschäftsbericht für das Verwaltungsjahr 1930, p. 78; Europäischer Fernsprech Dienst, Special No. 1930, pp. 18, 24, 26.

[27] It is quite difficult to follow Thomas' description of a totally new structure in preparing for war: "The existing star-shaped underground cable network had to be changed into mesh form and to be extented into rural areas in western and southern Germany," Thomas, Politics, p. 199, while comparing the map for the year 1943 (ibid., p. 200) with the older ones.

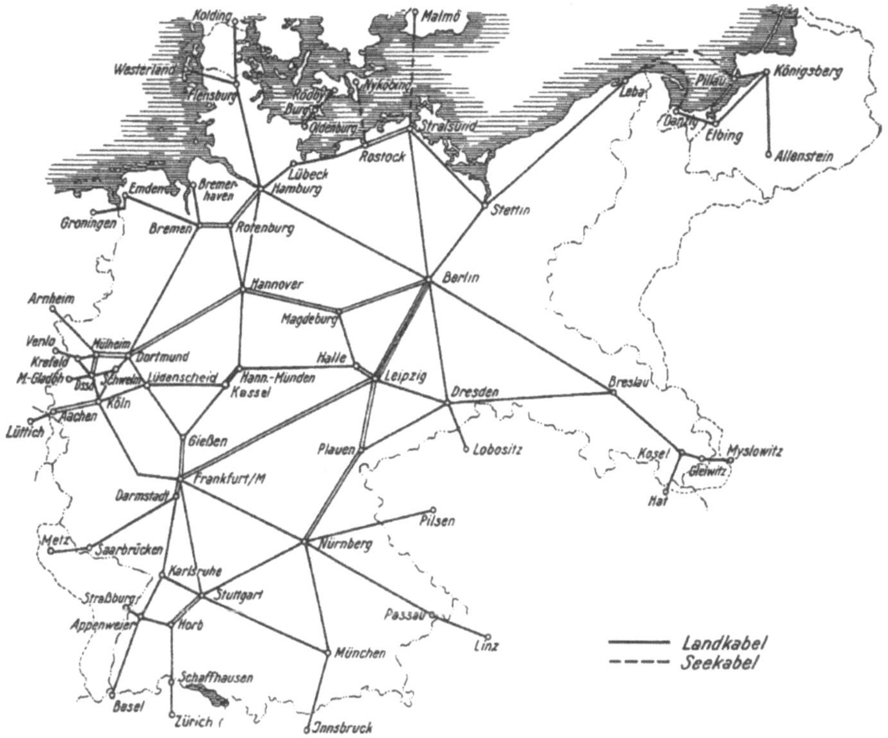

Figure 2: German long distance cable network, 1931
Quelle: EFD 28, 1932, p. 160

3. The Momentum

We need a broader definition of momentum than suggested by Hughes in his model, for while he stressed the specific technology, we have to include the preconditions for it as well. Particularly important here is the impact of the political sphere. Therefore, in momentum, we include all influences between the LTS and the outside world. Besides the inertia of the whole mass of the system in motion and its specifications of technology, it includes the initial and the later directions given from outside, the frames of development as well as the system's effects on the outside world.

Before the First World War the Deutsche Reichspost had decided to build an underground cable network. Surprisingly, soon after the German defeat, in February 1921, the Reichspost decided on a five year plan for the construction of a basic network. 5600 km of lines were planned[28]. In the following years the plan was not only realized, but greatly exceeded. The speed of construction was considerable, – more than 1000 km annually. The first five year plan for the construction of an underground cable network in Germany was implemented between 1876 and 1881. During that period more than 5000 km of telegraph cable were laid[29]. But both the actors and the legal framework had changed since then. In the 1870s the Reichspost took the initiative. The legal basis was weak, but fortunately nobody intervened. In the 1920s the legal framework was much stricter, but the Reichspost was only one of several corporate actors[30]. Other actors exerted considerable influence, including a number of German enterprises, cooperating with the Reichspost in the Deutsche Fernkabelgesellschaft[31], and the government itself was also involved.

Hartmut Petzold in his contribution titled "German-French Rivalry and cooperation in the construction of the European Telephone network after World War I" has shown the influence of the German government in the timing and speed of the growth of the network[32]. Of course the Reichspost was prepared to build and to run the network an underground cable network was thought to be safer than other alternatives and to have lower operating and maintenance costs all of which proved to be correct[33]. The industry, too, was eager to get big orders for construction as well as to have reference cables for export promotion. At that time, as a result of the conditions of the postwar era, the Reichspost was weakened financially by running deficits for the first time in its history. It had to be subsidized by the government. But apart from this one financial point, political considerations weighed most heavily.

---

[28] For further detail see Petzold, p. 272 ff.

[29] Cf. M. Geistbeck, Der Weltverkehr, Freiburg 1895, (repr.) Freiburg 1986, p. 481.

[30] Cf. in detail: Thomas, Politics, p. 182ff; idem, Korporative Akteure und die Entwicklung des Telefonsystems in Deutschland 1877 bis 1845, in: Technikgeschichte, 56, 1989, pp. 39–65.

[31] The ownership was shared between the Reichspost, which was the largest shareholder, and several enterprises; Petzold p. 273.

[32] There had been little initiative in this field from the French government, as Petzold shows. But there was too little technical as well as organizational competence and vision on the French side to match the German drive for either cooperation rivalry to occur. Therefore, the Petzold's approach seems to be a little arbitrary.

[33] In 1924 the annual report of the Reichspost mentioned that expected economic consideration had materialized "in their whole extent" (Geschäftbericht, p. 40). The safety of the system was proven during the winter of 1929. Again, as in 1909, when Berlin was cut off, thousands of masts for wires were broken by storms, but the underground cable network was not affected and communication was maintained.

The type of foreign policy for which the German foreign minister Gustav Stresemann later stood, was to be found in the field of international communications much earlier. In 1926, Stresemann pointed out that Germany should use its economic means in order to rebuild its reputation and its strength. Stress was placed on the international aspects of the 1921 decision to construct an underground cable network, as well as its development inside Germany. By offering excellent facilities for interstate communication, this traffic could be concentrated on Germany, which entailed the recognition of Germany as a partner on equal terms[34]. In the years following the Versailles Treaty this was a core aim not only of German foreign policy but of individuals as well. National pride had suffered. Though not openly stated, everybody realised that if European traffic was concentrated on Germany, this could help to bring the country back into the international community, at least in this particular field. The geographical situation in the middle of Europe should be exploited to its advantage. Cables offering connections for crossing state borders were built in preference to others[35]. By using advanced and reliable technology Germany would attract the traffic which was bound to emerge. For such a "pull-effect" an advanced cable network was indispensable[36]. The need to find work for to the masses of demobilised soldiers and unemployed, and to have reference points for export promotion, may have helped as well[37].

These calculations turned out to be right. From 1922 onwards the Reichspost received more and more applications from their counterparts in surrounding states for connections with the German cable network. The first interstate connection via Germany was opened on 1st May 1923, between Copenhagen in Denmark and Amsterdam, capital of the Netherlands. Applicants usually had to wait only a few months, during which the connection was tested under various conditions. But sometimes they had to "queue" several years, to be admitted. Normally the basis for the deci-

---

[34] Quotations from P. Craemer, at that time in charge of the development of the cable network, are to be found in Petzold, p. 271 f.

[35] "Das Landnetz mußte im Innenaufbau zunächst bis zu den Grenzen vorgetrieben und so eine ansaugende Kraft auf die umliegenden Länder wirksam werden, damit es seiner Bestimmung (sic!), als Zentralweiche des europäischen Fernsprechverkehrs zu dienen, gerecht werden konnte." (P. Craemer – see footnote above, quoted after Petzold, p. 272).

[36] Deutsche Reichspost, Geschäftsbericht über das Wirtschaftsjahr 1924, p. 39 f.; Petzold, p. 272.

[37] These additional factors are still to be explored. Only Petzold has made a step in this direction.

sion was not political, but technical. Connections were granted as soon as it was technically possible. In most cases this was based entirely on the extension of the cable network[38]. The existing network, which relied on surface wires on masts was not good enough for long distance calls.

Through the Reichspost's policy Germany was not only incorporated into the international European telephone network, but it attracted most interstate communication. Of course its geographical situation was of great help (see map of cables in Europe), but as the cases of other states show, it was not just this factor but the technical ability to offer quick and secure communication which was the decisive factor in attracting foreign traffic[39].

The political factor was also very important especially in the years after the First World War. When France and Belgium invaded the Ruhr district, all German applications for the extension of the cable network in the occupied area were turned down and further work was blocked[40]. On the other hand the German Reichspost blocked Belgian and French applications for communication with other states neighbouring Germany[41]. This embargo was lifted after the foreign troops were withdrawn. Belgium and France pursued different policies with respect to European communications. Belgium tried to obtain connections via Germany, and in May 1925 it applied for direct communication with Denmark. The Reichspost agreed in June 1925[42]. The time which elapsed between the application and the granting of the connection was used for technical optimisation. Austria, Czechoslovakia, and Sweden had asked for a connection with France via Germany, and the Reichspost had agreed. However, the French answer was still awaited[43]. The following year the French position changed. The reasons for this are not totally clear, but several factors might have played a role: the push of French businesses to communicate with their German

---

38 A. Zimmer, Über den Fernsprechverkehr fremder Länder im Durchgang durch Deutschland, in: Archiv für Post und Telegraphie, 1926, pp. 13–21.

39 The Hungarian post officer (Postrat), Dr. Havas, explained that with competing telephone lines in the Balkans the technically better cable connection could always attract most of the traffic. Cf. F. Havas, Der Kampf um den Fernsprech-Durchgangsverkehr, in: Europäischer Fernsprech Dienst 33, 1933, pp. 178–80.

40 Cf. Zimmer, p. 14; E. Horstmann, 75 Jahre Fernsprecher in Deutschland 1877–1952, Berlin 1952, p. 306; Petzold, p. 264.

41 Cf. Zimmer, p. 14 ff.

42 Cf. ibid., p. 14.

43 Cf. ibid., p. 15 f.

counterparts; the pull of the German cable network; the fear of loosing trade, since in 1925 Czechoslovakia, Denmark and Switzerland already had five international connections via Germany; and last but not least the political detente between Germany and France. The German performance in the international telephone organisation (CCI) was also a factor.

The CCI (Comité Consultatif International des Communications Téléphoniques à Grande Distance) met for the first time in April 1924 in Paris to discuss technical matters related to the European telephone service. It is worth noting that Germany was invited to the meeting in Paris with the same rights as all other nations as early as 1923, a time of extreme political tension. Because of their experience managers of AT&T also attended.

A considerable number of problems had to be solved, since each country had its own standards. The technical systems were different, and the material used varied as much as the organisations. In 1922 AT&T had suggested the formation of a single supernational European telephone company on a private basis for all countries[44]. This suggestion which called for far reaching integration was discussed, but it never had a chance to materialize for nationalistic reasons. The CCI worked sucessfully because it merely suggested technical standards and solutions to be decided upon in each state individually. The CCI met annually or every second year, while in the meantime special technical committees worked on specific problems[45].

Inside these committees Germany could exert considerable influence. It had, apart from the US, the greatest experience of running the biggest cable network. In matters of telephone technology it was matched only by the USA and by Sweden, and in cable production only by the United States and by Great Britain[46]. Furthermore, Siemens & Halske was the only European firm to produce in-house both cables and telephone technology. Experience and technical superiority made it comparatively easy for the German representatives in these committees to press for German standards in technical matters. In these areas nationalistic prejudices could

[44] Ibid., p. 276.

[45] All committees and their topics are listed in: Valensi/Georges, Die fünf ersten Jahre des Zwischenstaatlichen Beratenden Ausschusses für den Fernsprechweitverkehr (CCI), in: EFD 12/13, 1929, pp. 103–17.

[46] For the international competition on the world telephone market see: A. Attman/J. Kuuse/U. Olson, 100 Years LM Ericsson, Vol. 1, The Pioneering Years, Struggle for Concessions, Crisis 1876–1932, Stockholm 1976; V. Schröter, Die deutsche Industrie auf dem Weltmarkt, Frankfurt 1984, pp. 331–41.

not prevail. Thus the momentum of the German cable system caused a spillover effect on the development of the European system. In the technical field the integration of Europe was far ahead of its political institutions.

Another factor which promoted the momentum of the system was the earning of foreign exchange. In 1925, when communication with foreign countries was still in its infancy, Germany earned about 3 million Gold-Franks in foreign exchange from its services[47]. Later, when traffic had grown in extent, this factor became quite important, especially against the background of the World Economic Crisis in the 1930s, when foreign exchange was short in most parts of Europe. Charges for service were agreed upon internationally[48], but several countries started to compete with each other by means of undercutting the internationally laid down charges in order to attract foreign interstate calls[49].

The momentum of a system is not only based on sums already invested or even once and for all decisions. The common view taken of it by all the people confronted with it, and especially by its workforce is also of importance. If this corporate identity, which gives meaning to and for action, is broken, the momentum, even of a big LTS, is in jeopardy. It is worthwhile, therefore, to consider how German technicians perceived their work in an international context. At this level two points were stressed again and again in various contributions to technical periodicals. The political one focussed on a way of overcoming the underdog complex which had emerged from the lost war and the Versailles Treaty. In this context it was a matter of pride to develop the most advanced technology, primarily from German sources[50]. The other point was similar, but not so outspokenly political. At that time it was already deeply rooted inside German engineers that German technology should not be cheap but should be the best[51]. This common standard was set before the turn of

---

[47] It was agreed that all international payments had to be calculated in a theoretical currency called "Gold Frank" (GF), shaped according to the Swiss Franc, but based entirely on the price of gold. In 1925 German average expenses for calls abroad totalled 382,000 GF per month, while service for foreign calls earned 630,000 GF. Cf. Zimmer, p. 13.

[48] On the calculation of charges see K. Ehlers, Deutschland als Durchgangsland für den zwischenstaatlichen Fernsprechdienst, in: Jahrbuch des elektrischen Fernmeldewesens, 1938, pp. 324–40.

[49] Cf. Ehlers, p. 335; Havas, p. 179 f.

[50] E.g. Ehlers and various contributions in EFD and Archiv für Post und Telegraphie.

[51] Cf. e. g. Sautters, p. 258; viz. preceeding footnote.

the century[52] and still has its repercussions today.[53] It was understood that the Reichspost should be prepared to meet the demand for communication from other nations, because the geographical situation of Germany "naturally" imposed this "duty" on it[54]. Or as a leading manager expressed it: "Right from the beginning of the construction of the cable network the RPM (Reichs-Post-Ministerium) stressed that the highest possible technical standard should be used, as the German cable network would be charged with extremely important duties for European traffic, and therefore it should always be prepared to meet the highest demand"[55].

4. *Reverse Salients*

In such a context of highly valued technical perfection, reverse salients were extremely influential. In many cases it was not private industry, driven by wishes for more profit, that pressed for one innovation after the other, but the Reichspost itself. On various occasions the Reichspost encouraged industry to take up research in an important new field of technology[56]. Moreover, it kept its own an organisation for research and development. The first laboratory was founded in 1888, as "Telegraphen-ingenieurbureau des Reichspostamtes", it was extended and renamed "Telegraphenversuchsamt" in 1899, and consisted of decentralized laboratories in different cities. In 1920 they were unified again in the "Telegraphentechnische Reichsamt", which from 1928 onwards was called "Reichspostzentralamt". The Reichspostzentralamt was located in Berlin, but gradually it built up sub-laboratories in different towns. This apparatus worked parallel to R&D departments in industry, with which there was considerable interaction[57].

---

[52] U. Wengenroth has shown that the standards for German goods were different in the second half of the last century. Cf. German steel: bad and cheap, in: Technikgeschichte 54, 1987, pp. 197–208; Wengenroth, Unternehmensstrategien und technischer Fortschritt, Die deutsche und die britische Stahlindustrie 1865–1895, Göttingen/Zürich 1986.

[53] Today some authors still put Germany in a "techno-perfectionist" context in which "overengineering" is at hand (see R. Mayntz/V. Schneider, The Dynamics of System Development in a Comparative Perspective: Interactive Videotex in Germany, France and Britain, in: R. Mayntz/T. Hughes, pp. 263–98, p. 291.

[54] E.g. Ehlers, p. 325.

[55] Sautter, p. 259.

[56] Petzold gives an example on the automatic telephone exchange, p. 268.

[57] Cf. H. Kröber/K. Heyer, Geschichte der Leitungs- und Übertragungstechnik, in: Archiv für Postgeschichte 1, 1983, pp. 130–45 (part 1) and No. 2, 1983, pp. 112–28 (part 2).

In order to improve facilities the Reichspost payed special attention not only when a cable line was constructed, but to its maintenance as well. As in other LTS, a specific group of personnel were trained to permanently oversee the system. In 1922 commissionaries (Fernleitungskommissare) were attached to each major area (Oberpostdirektion) with the task of spotting weak points in the system and to work for their elimination[58]. In the light of Hughes' model these Fernleitungskommissare represented an organizational guarantee of the perpetual promotion of reverse salients.

Some examples of important reverse salients are worthy of mention. Pupin coils and amplifier valves were the precondition for long distance communication. They too were improved, but the major issues were standardisation, automatisation, carrier technique and improvements in transmission. When in the early 1920s tests showed that a four-wire connection was much better than a two-wire one, all long distance calls were run on this standard. This decision demanded a certain type of cable, in which two and four wire connections were combined. These cables had three fold insulation: first each wire, second all wires belonging to one connection separately, and third all groups of four against all groups of two wires. The standard German cable was more complicated, thought to be more reliable, and looked different from standard cables of other nations[59].

Another improvement representing the influence of reverse salients was the echo suppressor, which was introduced from 1925 onwards. The echo effect caused considerable problems, because part of the transmitting wave was reflected (and re-reflected) at the receiver, which meant that the speaker was forced to listen to what he had said before, whilst still speaking. The echo suppressor was a major step towards understanding during long distance calls.

The construction of direct lines between different states running through Germany become very important. They were laid according to German standards by the Deutsche Fernkabelgesellschaft and maintained by the Reichspost, but no exchange, switching or anything else was done in Germany. In 1931 there were 23 of these direct lines, while their number had reached 86 in 1939[60]. These direct lines were leased by neighbouring states for their own traffic. Switzerland even leased a line, not for

[58] Cf. Horstmann, p. 306.

[59] Cf. C. Jacobaeus, 100 Years LM Ericsson, Vol 3: Evolution of technology 1876–1976, p. 199; for the technical standards of the German long distance cable see Sautter, p. 258.

[60] Cf. Deutsche Reichspost, Geschäftsbericht für das Rechnungsjahr 1930, p. 82; Ehlers, p. 333.

its own use, but for the sub-service for Italian-British communication. The direct lines showed clearly that the international orientation of the Reichspost was liberal, aimed more at technical possibilities than at political discrimination.

Other improvements were not completed in the interwar period. By applying carrier technology it was possible to allow up to 12 calls on one pair of wires simultaneously. This was a great advantage, but when, in the early 1930s, the Bell Laboratories in the US invented a new type of amplifier valve, it was possible to have 200 calls! For this a new type of cable, the coaxial cable, was needed. One year after its invention in the United States, technical work on this was started in Germany, and in 1934, the first coaxial cable lines were constructed. From May 1938 onwards such a line was in use between Berlin and Munich. In 1941 3,900 km cable lines of this type were in use[61]. Other nations prominent in telephony, such as Sweden, only installed their first coaxial cables after the Second World War[62].

The second improvement which took decades to implement completely, was automation. With the rising amount of telephone calls the need for automation became pressing. The connection time for long distance was considerable, because several operators had to cooperate. At each level of the hierarchy from the telephone receiver to the local switchboard to the exchange for the area and then on to that for long distance calls, and down the same ladder again at the other end, an operator had to be told what to do, which took much time. If the person at the other end was not at home, there was no charge for the service, nor for the use of the line, which often lasted for several minutes[63]. Furthermore, in rural areas the telephone exchange was open only during office-hours, which meant that at other times there were no calls, except if the exchange was automated. Understandably, demand for quick automatic connection was great. There were only three enterprises in the world able to produce automatic exchanges, and one of them was German[64]. The Reichspost invested so heavily in automation that it took a world lead in this field. In 1938 88.5% of all telephone sets were connected with automatic exchanges[65]. For technical reasons long distance calls were the last to be-

---

[61] Cf. Sautter, p. 260.

[62] Cf. Jacobaeus, p. 198.

[63] Cf. Zimmer, 1926, p. 20.

[64] Headed by AT&T, LM Ericsson and Siemens & Halske.

[65] Corresponding figures for other nations for that year were: Switzerland: 86.0%; USA: 58.0%; Sweden: 46.1% (Horstmann, p. 316). Automatisation of German local nets covered 14% in 1925, 57% in 1930 and 83% in 1935 (Sautter, p. 249).

come automated, but in 1938 a substantial number of German telephone sets could be reached from a single operator, which meant that the above mentioned ladder of hierarchies had to be climbed, but no longer decended step by step[66].

Besides the major efforts which have been mentioned, an unknown number of smaller improvements were tested and put into operation, every case representing the effects of reverse salients. It seems that financial considerations were placed second to technical ones. The aim of the early 1920s, to build an underground cable network, and to maintain it in the vanguard of technical development, was reached. Incentives based on reverse salients worked extremely well within the German cable system. All of this attracted foreign traffic from other European countries, which can be taken as a sign of integration.

5. *Load Factor*

The load factor is defined by the ratio of the average to the maximum use of the system. At first glance the load factor is a strong argument because it is easily accountable. But what an agreeable "average" is, has to be defined as does the "maximum". When there is a load which is changing over time, it has to be defined in a certain way. The load factor has to be calculated for the hours of peak demand[67], so the figures published by the Reichspost for daily or annual use are not suffficient. Within the cable network only workdays (six per week) were counted for calculation, and as many exchanges were manned only during working hours, all the calls were expected to be exchanged during this time. On this basis the CCI suggested 200 calls per workday, with an average duration of three minutes, as the maximum load of a line. More calls gave an incentive for the construction of a second parallel connection. 200 calls lasting three minutes did not add up to 600 minutes or ten hours time for the cable to be employed, as the connection had to be established first and, after termination, to be disconnected for the next caller. The direct lines mentioned above helped to reduce this time. Before they were constructed, such direct lines were switched together for a certain time of the day; e. g. direct traffic on such a basis was permitted in February 1924 between Oslo and Copenhagen on one side and Amsterdam/Rotterdam on the other from 10 to 10.30 a.m., 4 and 4.30 p.m., and 9 p.m. until 8 a.m[68].

---

66 Cf. Horstmann, p. 316.

67 As it is done for electrcity supply (see Hughes).

68 Cf. Zimmer, p. 14.

In order to obtain a higher load, important telephone exchanges were manned for 24 hours. Those hours with low traffic were offered for foreign use, as illustrated in the Scandinavian – Dutch case mentioned above. Later, when the capacity for more traffic was constructed, foreign calls passing through Germany were admitted without any restriction.

The demand for long distance calls was rather steady. The number of minutes required was 887 million in 1923 and 813 in 1927[69]. The number of all calls (including international ones) amounted, in 1928, to 270 and in 1937 to 282 millions[70]. These figures do not show a steeply rising demand, which could have prompted further extension of the cable network. The load of the system alone can only explain the heavy investment up to the year 1925, when demand was met. This meant that a line for communication could be given at the time the customer wanted it. With the extension of the cable network, he did not have to wait for hours until a free line was allocated to him. Before 1925 bottlenecks in demand were reflected in the requirement for "urgent" calls, which were charged threefold. This dropped heavily from 26.5% of all long distance calls in 1923 to 1.7% in 1927[71]. From 1925 onwards customers could call when they wanted to do so. After that year the load factor could not have played a major role in the extension of the network.

The structure of charges was for the load factor. Especially to start with, when there was too little accounting, the charges were settled quite arbitrarily for calls inside Germany only[72]. Charges for foreign calls or those passing through Germany were charged according to contracts, which were based on the length of the lines. No political preference or discrimination were to be found in these charges, a fact worth noting, not only because discrimination could be expected after the First World War, but this is also important in relation to the question of European integration. The fee for a call from Brussels to Copenhagen was fixed at the price agreed upon in 1903, more than 20 years before the line became operational[73]. Charges for the Austrians, being former allies, for the Swiss,

---

[69] Additional figures are: 1924: 792; 1925: 830; 1926: 775 million minutes (all figures calculated from Deutsche Reichspost, Geschäftsbericht über das Rechnungsjahr 1927, p. 63).

[70] From 1928 onwards the number of calls was counted and no longer the minutes. (Deutsche Reichspost, Geschäftsbericht über das Rechnungsjahr 1929, p. 62; EFD 47, 1937, p. 279).

[71] Corresponding figures are for 1924: 8.8%; 1925: 4.1%; 1926: 2.2% (Deutsche Reichspost, Geschäftsbericht über das Rechnungsjahr 1927, p. 63).

[72] Cf. Thomas, Politics, p. 195.

[73] Cf. Zimmer, p. 14.

being neutrals and for the Belgians as former enemies in their calls to Danzig were all the same. The fee for the standard call (three minutes, no special service for urgency etc.) from Copenhagen to Vienna was 8.95 Gold-Franks (GF) in 1925. From that sum Denmark received 3.00 GF, Germany 3.30 GF, Czechoslovakia 1.50 GF and Austria 1.15 GF[74].

These standard charges could be reduced by various means including night calls and fixed times, which were offered in order to spread load from peak demand hours over the whole day and night. Other reductions were not aimed at the spread of demand, but at the attraction of additional traffic. From the years of the World Economic Crisis onwards, several nations competed with each other by lowering their charges for through calls, but in most cases the undercut state lowered its charges as well[75]. In the end – as previously – the customer chose the best technology at the same price. This choice worked as an incentive to extend the underground cable network. In the end experience showed that by lowering charges very little, international traffic could be raised[76]. A special investigation showed that repercussions from the World Economic Crisis were felt with a delay of three years[77].

There is no doubt, that in any telephone system the majority of calls are local, and the majority of long distance calls are national. But the actual ratios between long distance and international calls are to a certain extent dependent on the size of the state and its market. In 1930 international calls represented only 2.4% of calls in Germany but 28.0% in Hungary, while the other European nations were situated between these levels (see Table 1). But the number of calls crossing the German border was much bigger than in the case of Czechoslovakia. However, calls crossing a country were the smallest number of all. For Germany, the state through which most of these calls were sent, their number was below 100,000 in the 1920s, climbing up to a zenith of 124,000 in 1933, and fluctuating thereafter[78]. The use of direct lines passing through Germany needs to be added to this, as these calls were not counted[79].

---

[74] All charges for the early 1920s are listed in: ibid.

[75] Cf. Havas.

[76] Cf. Ehlers, p. 330.

[77] No explanation was given for this fact, shown for several European states. (anon., Die Wirkung der Krise auf den zwischenstaatlichen Fernsprechverkehr von 1932 bis 1934, in: EFD 42, 1936, pp. 34–36.

[78] Number (in thousands) of international calls passing through Germany with service of a German operator: 1933: 124; 1934: 116; 1935: 84; 1936: 93; 1937: 96 (Ehlers, p. 334).

[79] According to Ehlers' figures (p. 332 f.), these figures were much higher. Based on an estimated length of six minutes we can calculate 750,000 calls, which appears to be too high.

Table 1: International calls in total numbers and in % of national long distance calls

State	1931 int. calls, No. in mill.	1931 %	1937 int. calls, No. in mill.	1937 %
Austria	3,5	22	1,6	21,7
Belgium	3,1	6	3,2	7,9
Bulgaria	0,03	0,3	0,08	2,5
Czechoslov.	2,7	16	1,8	10,1
Denmark	0,8	1	0,8	0,9
Danzig	0,8	53	0,6	51
Eire	0,3	13	0,4	13,2
Estonia	0,08	3	0,09	2,6
Finland	0,2	2	0,2	0,4
France	3,4	2	*	*
Germany	6,8	2,4	4,3	1,5
Greece	0,001	0,2	0,05	2,8
Hungary	1,6	28	1,2	24
Italy	1,3	3,8	1,7	5,3
Latvia	0,2	2	0,2	1
Lithuania	0,3	10	0,09	4,9
Luxemburg	0,7	18	0,4	11,5
Netherlands	2,3	9	2,1	5,2
Norway	0,3	2	0,4	2,3
Poland	2,3	8	1,5	5,8
Portugal	0,05	1,4	0,07	1
Romania	*	*	0,4	5,9
Spain	0,3	2,2	0,2	1,1
Soviet Union	0,1	0,4	*	*
Sweden	1	2	1	1,9
Switzerland	4,6	6	4,5	4,7
U. K.	1,1	1	1,7	1,7
Total Europe	37,861		28,58	

* No data available
Source: EFD 30, 1930, tables p. 268–275; EFD 47, 1937, tables p. 278–286.

Viewed only from the load figures the international extension of the cable network, and with it the integration of Europe, were nothing more than an appendix. Given this background the attention paid to it by managers and technicians of the LTS is astonishing. The special care with which international calls were treated in the LTS shows that the load factor in the cable network can tell us much more about national needs than about international requirements.

6. The Extension of the German Telephone Cable Network and its Repercussion on European Integration

With the construction of the German cable network the technical means for the integration of Europe were enlarged and improved yearly. Not only was Germany technically integrated, but via its cable network neighbouring countries were as well. The German government, which in 1920/21 had pressed for the construction of the cable network, did not intend to promote European integration, but its construction obviously proved to be an unintended step towards it. The cable network in itself did not represent integration but rather was a precondition for it. Any possible integration in this context is to be understood as an unintentional side effect of mass use of the cable network. Private customers made very few long distance or even international calls. The two most important groups of customers were industry, including the third sector, and within it, news agencies and journalists especially. Contemporaries took the amount of telephone communication as a correlation of economic relations between the countries involved[80].

It can be seen from various tables that up to the World Economic Crisis demand for international telephone communication rose[81]. Unfortunately there is a break in the basis of calculations. Up to 1930 minutes were counted, while from that year on it was the number of calls. But table 1 shows a sharp decline in the absolute number of international calls for nearly all European states. From this development alone it is not to be concluded that integration decreased, since all economic activities slackened. But in most states the percentage of international calls among all long distance calls fell; an observation which was especially valid for those nations with the greatest amount of long distance calls, such as Germany and Switzerland.

The technical means represented by the cable network helped with the integration of Europe without intending to. This development, lasted throughout the 1920s, until governments thought that the World Economic Crisis could be stopped by heavy political and economic interference.

7. Conclusion

Our questions were approached by applying Hughes' model of Large Technical Systems. Though the model was constructed for a whole LTS,

[80] Cf. Horstmann, p. 310; Ehlers, p. 330.

[81] Annual tables in EFD up to 1930, based on the amount of minutes, not calls.

the application showed that it is useful for the investigation of parts of a system, as long as these parts form a subsystem with defined borders of their own. It identified areas for investigation and helped in understanding what made the German cable network expand continually.

Demand or load factor propelled expansion only during the first years, up to 1925. But momentum and reverse salients were important enough to cause further growth of the system. For the momentum several factors played a role, which in the course of time varied in their importance. In the early 1920s foreign political considerations, especially a desire to bring Germany back into the community of respected states, played a bigger role than the earning of foreign exchange. During the 1930s this was reversed. Important, too, were the beliefs of technicians, engineers and bureaucrats inside the system, who sought to offer the most advanced technology. Costs played a secondary role, and with the exception of the early 1920s, the Reichspost was profitable and its considerable earnings were reinvested. In this atmosphere reverse salients propelled the system, especially as inventions were not only offered by industry from outside the system. Industry acted as a corporate player by being, through share holding, part of the cable network itself, and the Reichspost itself formed organisations for the promotion of the technical and organisational progress of the network.

The German cable network with its high quality technical development and its many international connections represented a great opportunity for the integration of Europe. For the first time in history immediate and uncomplicated communication not only to and from Germany, but throughout Europe was at hand, offered at much cheaper rates than those of the telegraph. Technical cooperation provided by the European CCI worked smoothly. Technical preconditions for integration were improved every year. It is disappointing to realize the small extent to which this offer was taken up. While progress in integration was to be seen during the late 1920s, the second decade of the interwar period is better characterized by disintegration. The developments of the 1930s led to divergence, not to any further convergence. While communications engineers worked successfully for technical progress, their results were demanded less. Advanced technologies for communication proved neither to be a glue for the intergration of Europe nor a means of reducing the room for hostile political manoeuvre between different European states.

II.
The Transfer of Technology and its Diffusion

Introduction

G. NICK VON TUNZELMANN

Perhaps the most important result to come out of the substantial body of recent theoretical and empirical work on technology transfer has been built on the proposition that such transfer may involve substantial commitments on the part of the company or country to which the technology is being transferred. The commitments may amount to no more than monetary sums for the purchase of "turnkey" plants, but even here the sums involved, aside from the cost of the licence itself, often represent a considerable fraction of the initial innovation costs[1]. However such turnkey transfers have been frequently criticized for perpetuating technological dependency by the borrower; for example, in postwar Latin America[2]. Even turnkey plants require major learning processes for the borrower in terms of operating functions, but the object of technology transfer is usually to transplant capabilities not only in operation but also in investment and ultimately innovation. The borrower companies or countries want to be able to overcome technological dependency, and establish their own capabilities for setting up new plants and equipment, and eventually plants and machinery that improve upon the standardized design first borrowed. The learning processes involved in these stages involve high degrees of commitment of people as well as finance. This arises because technologies borrowed rarely are confined entirely to blueprints, or formal documentation: most technology involves considerable and often predominant areas of tacit and uncodified knowledge, which the lender does not incorporate into the package (and normally is unaware of its explicit

[1] Cf. D. Teece, Technology transfer by multinational firms: the resource cost of transferring technological know-how, in: Economic Journal 87, 1977, pp. 242–61.

[2] Cf. J. M. Katz (ed.), Technology Generation in Latin American Manufacturing Industries, London 1987; C. Freeman, Catching up in world growth and world trade, mimeo, SPRU, Sussex 1992.

content), and the borrower knows neither its content nor its role[3]. Thus technology cannot usually be regarded as simply "information" in the usual economist's sense, because it incorporates such uncodified (and unpriced) information; conversely, for the borrower, the learning process is not just the relatively costless process implied in traditional models of "learning by doing" (where it is a by-product of operating functions), but a sustained and often costly involvement of people and funds. Included in the recent empirical results have been extensive support for the importance of formal infrastructural investment in education in the countries where transfers of heavy and advanced industries appear to have been most successful in the past two decades, including the importance of producing or attracting large numbers of qualified scientists and engineers[4]. Similarly, Japan moved rapidly to establish both formal training for engineers and extensive practical work in the course of their training[5].

These rather general results from recent technology transfer put the discussion of the two rather specific papers under discussion into a broader context. For this, we additionally need to unearth more not only about the innovation process in the leading-edge companies or countries, but also about the "imitation" process in the followers (bearing in mind what was implied above about the local improvements in even imitated technology by the borrowers). I have elsewhere developed taxonomies for representing industrial production in firms or countries, and utilized it for both historical and present-day analysis[6]. In this perspective, Braun and Edgerton's paper particularly emphasizes the performers – military vs civil. Sally Horrocks's instead considers innovation in products vs processes, and also relates to scale and scope issues. In both cases I would like to add a missing dimension to their work, though principally I shall embroider the work they offer. I should begin by stressing that, in their own terms, both are highly professional and illuminating pieces of work, and my comments assume this to be so. I therefore see my role as being one of enlarging upon them both.

---

[3] Cf. N. Rosenberg/C. Frischtak (eds.), International Technology Transfer: Concepts, Measurements and Comparisons, London 1985.

[4] Cf. e. g. L. E. Westphal/L. Kim/C. J. Dahlman, Reflections on the Republic of Koreas Acquisition of Technological Capability, in: ibid., pp. 167–221.

[5] Cf. H. Gospel (ed.), Industry Training and Technological Innovation, London 1991, chs 5–7.

[6] Cf. N. von Tunzelmann, Technology and Organization during the Industrial Revolution, in: P O'Brien/R Quinault (eds.), The Industrial Revolution and British Society, Cambridge 1993, pp. 254–82; N. von Tunzelmann, The supply side: Technology and History, in: B. Carlsson (ed.), Industrial Dynamics, Boston 1989, pp. 55–84.

As applied to the Braun/Edgerton paper, the taxonomy I have described might classify the topics raised in the following way:

Technologies	Processes	Products	Performers
Superchargers	Materials	High-speed	Military/civil
Engines	Mass production	High-altitude	public/private
Science (aerodyns.)		War-directed	Sector spinoffs
R&D programmes		Standardized	

We may usefully concentrate on the final column of performers, since this raises the issue of technology transfer most directly. On p. 121, the authors appear to finesse the most common question, of spillover from military to civil use, by declaring that the industry was really a military-oriented one, even in peacetime, in the sense that even the civil aircraft industry was probably not a part of the civil economy. Looked at from my standpoint, it seems as if they have some justification for such a contention in terms of the supply side – the technologies, processes and firms involved were similar for civil and military aircraft. But this is less true of the products. The paper refers to some of the criteria for successful military aircraft – others may be added, such as bomb payload. These differ in many cases from the criteria for successful civilian aircraft, which would include passenger density, length of haulage, fuel economy, and similar measures. It would seem unfortunate to eliminate the most obvious question by assumption at the beginning, and our separation between supply and demand aspects shows that this is unnecessary. The military/civil question now becomes: even if the firms (etc.) are the same in the two, was the progress on the civil side faster or slower than would have been the case without the military side?

An answer to this question is, of course, far from easy, because counter-factuals are being explicitly posed. However it is not difficult to show that counter-factuals are at least implicitly posed for any issue in this or many other areas. The counter-factuals here differ from those given most weight in the well-known quantitative studies of economic history of transport, like Robert Fogel's on the US railroads[7], in that they are counter-factuals developed over (continuous) time. The study by Fogel and others in similar vein, much influenced by received wisdom in neoclassical economics, instead conduct the counter-factual at an end-point in time, ie, as a comparative static exercise. Fogel himself is too good an historian to be taken

---

[7] Cf. R. W. Fogel, Railroads and American Economic Growth: Essays in Econometric History, Baltimore 1964.

in by this, and gives some concession to a more dynamic technological model in which the internal combustion engine is invented earlier in the hypothetical absence of railroads.

The fact that counter-factuals are implied is not necessarily as fearsome as some may think. What is being requested is, on the one side, a "factual" study of the technological trajectory pursued by civil aircraft, mapped in terms of the criteria thought to be relevant to such a field. On this, we have considerable historical evidence already, for example explaining the contribution of the Douglas DC3 or more recently the Boeing 707. Of course, problems may be encountered in projecting this information back from the 1930s to the 1920s and before, but these are standard historical problems. Then we have to derive a counter-factual alternative, for which some linear or loglinear extrapolation may be required – here our historical background has to come into play, to justify whatever hypothetical alternative we choose to adopt. I sincerely believe that the conceptual problems are nowhere near as great as perhaps even my description makes them sound. However the data problems may (or may not) be considerable.

Casual inference suggests that the civil aircraft industry of 1920, for all the difficulties of postwar readjustment, would have been far short of its actual level without World War I, although this of course would need to be established. Assuming this were found to be true, one might then wish to examine the opportunity cost argument – would the civil branch have developed not just as rapidly but also at lower cost than through the actual experience of war-demand fostered military growth? It would be astonishing to find that the civil trajectory could not have been promulgated at lower cost than actually experienced, through all the waste of equipment and human lives, by wartime endeavour. But this simply shows that we are not posing the counter-factual correctly, for it is just as obvious that without a war, nothing like that quantity of resources would have been thrown at civil aircraft. At this point, the counter-factual does indeed become immensely difficult to appraise.

Historical comparisons give no simple guidance about the matter. On the one side, we have the Mary Kaldor argument about baroque military technologies, touched on in the Braun/Edgerton paper[8]. In Kaldor's view, military technologies in peacetime are baroque in two main senses: (i) they are over-ornate, because of "soft budget" constraints to spending more and more on product development, for example the $600 lavatory seats

[8] Cf. M. Kaldor, The Baroque Arsenal, London 1982.

that so offended President Reagan; (ii) they are outdated even before they are realized, in that they are almost invariably trying to find a better way of fighting the last war, and without the immediate competition of a new war they have no standard of acceptability. It should be noted that Mary Kaldor implies a comparison between military technologies in a context of war and the same in a context of peace. The present authors are well aware of this, but might perhaps express the point more clearly – in a dynamic sense (as implied by my previous paragraph, and also by the reservations expressed in the conclusion of their paper) one may not be able to draw hard-and-fast distinctions between wartime and peacetime.

The answers, such as they are, are by no means as self-evident as might be imagined. A good example is the history of microelectronics. Early development of integrated circuits was effectively sponsored by the NASA space programme. The need to cut weight and physical dimensionality to the minimum dictated an imperative towards miniaturization. As Swann and others have shown, this induced trajectory of rapid miniaturization happened to resolve at one and the same time nearly all the technical bottlenecks of the existing technology – further miniaturization resulted not only in greater component density (as the major gain) but also in higher speeds (from reduced distances), smaller heat dissipation, etc.[9]. The technological gains of the microelectronics industry after World War II were the most rapid sustained advances on record – essentially because the trajectory imparted by the needs of the space programme simultaneously resolved virtually all the practical shortcomings of the preceding technology even as viewed from the demand side. Such a happy conclusion has been rare in human history – the more common story has been one of trade-offs in technological performance whereby gains in one dimension have been offset by losses in another, or the gains on the supply side have been turned into losses on the demand side. In that sense, aeronautics has perhaps more often followed the traditional than the microelectronics path.

In regard to processes, my personal casual inference is that process technology rigidifies in war – the extremely urgent demands for war material necessitate the rapid acceleration of existing methods of production, rather than lateral investigation of new methods of production. Many were the entertaining stories told of this in World War II[10]. The

---

[9] Cf. G. M. P. Swann, Quality Innovation: An Economic Analysis of Rapid Improvements in Micro-Electronic Companies, London 1986.

[10] Cf. M. M. Postan, British War Production, London 1952.

authors might query the extent to which this war footing aided or perhaps constricted the post-war shift to civilian needs in aircraft production.

They are, in my view, correct to draw attention to the possible spillover in materials, promulgated by aircraft needs but with conceivable ramifications for other sectors. It is not only the direct impacts on other products like motor cars which are relevant here – there may again be upstream impacts on the rise of new sciences like materials science, and the development of new sectors like the plastics industry, which was the fastest-growing industry in Britain after World War II. At this point should be added the possible spillover of tools and measurement apparatus. The importance of instrumentation in spillover from science to technology has been recently argued by Rosenberg[11], and I would add to his work the possible spillover from technology to other technology.

An even broader process spillover suggests itself. It is well known that fast-growing sectors in the modern US economy are partly created by the mobility of skilled employees, who in some cases split off from their parent employer to form their own companies, in fields such as IT and biotechnology. Mobility of skilled employees has historically been a major source of diffusion across countries, a fact perhaps best demonstrated in Kristine Bruland's analysis of the use of British skilled workers in the creation of the Norwegian textile industry in the 19th century[12]. The point is made in one sentence of page 127 of the present paper, but it deserves more. As Bruland shows by implication, a transfer of skilled workers, even if long sustained and supported, does not necessarily result in successful transfer of technology.

Time does not permit me to examine all the functions implied in the Braun/Edgerton paper. I would however like to draw attention to one facet. It has been shown that the spin-off in terms of materials from the Zeppelins after World War I was much more dynamic in the USA than in the UK, when both countries became entitled to utilize the Zeppelin technology[13]. This was because in the US, the new materials were taken up by the materials sector, in firms like Alcoa, whereas in Britain they fell to the responsibility of the more stagnant military equipment sector, in firms like Vickers. This, together with some of the points made in the

---

[11] Cf. N. Rosenberg, Scientific instrumentation and university research, in: Research Policy, 21, 1992.

[12] Cf. K. Bruland, British Technology and European Industrialization: the Norwegian textile industry in the mid-nineteenth century, Cambridge 1989.

[13] Cf. M. Graham, R&D and competition in England and the United States: The Case of the Aluminium dirigible, in: Business History Review, 62, 1988, pp. 261–85.

paper, rather indicates that the technology transfer issue needs to be examined not just statically, in terms of which particular techniques etc. were transferred to other sectors, but dynamically, in the whole growth performance of those other sectors. At this point the measurement problems again become serious and perhaps intractable.

Horrocks in her paper makes valuable use of new source materials to re-evaluate R&D in British industry between the wars, and especially in the somewhat neglected food processing industry. She is right to correct the undue emphasis on a limited number of old and new industries. However the implication that some may unwittingly draw from her paper, that historians have tended to assume that the food-processing industry was technologically and economically stagnant, would not be correct. The food industry has traditionally appeared on even quite restricted lists of "new industries" of the interwar period, such as those compiled by Aldcroft[14]. The difficulty is that it has never been clear quite why it does appear, although its relatively rapid rate of growth presumably has something to do with it. A more constructive reason is that, at least in the 1920s, certain branches of food-processing were in the forefront of organizational change and thus the "rationalization movement", as Hannah has shown[15]. This was consummated in the giant mergers of the late 1920s, such as Unilever (which Horrocks appears to be excluding from foodstuffs) and those in brewing. Thus, utilising my taxonomy, it has been through calling attention to the organization function that the food industry has been reinstated among the progressives.

This raises the issue of how "new industries" (or old) can properly be defined. I have expanded at length on this topic elsewhere, so do not propose to do so again. Suffice it to say that I found in the literature at least nine different ways of implicitly defining new industries, of which the exploitation of new technologies was but one. Indeed new technologies by themselves turn out to be a limited guide to new industries, since most of the fundamental technological breakthroughs (here aircraft are a partial exception) had taken place some 40 to 60 years earlier. Each criterion – of which new forms of organization was another – generated a different list of "new industries", with by no means complete overlap with all the other lists. There was a particular disjunction between definitions which could be regarded as essentially supply-side ones, like new technologies

---

[14] Cf. D. H. Aldcroft, Economic growth in Britain in the Interwar Years: A Reassessment, in: Economic History Review, 20, 1967, p. 311.

[15] Cf. L. Hannah, The Rise of the Corporate Economy, London 1976.

or organisational forms, and those which were demand-oriented. The particular branches of food processing focussed on by Horrocks probably better fit demand-side interpretations of new industries than supply-side ones. The lack of attention by others to food may therefore partly reflect the fact that demand-side changes tend to be less revolutionary than some on the supply side.

It also partly reflects the fact that food processing produces a product that is a final consumer good. I guess one could devise a broader economic model in which food was an input to labour etc. ("A Mars a day helps you work, rest and play"), but the orthodox approach stops short of this. As a result, there is no downstream spin-off – product innovations in the chocolate industry do not generally become process changes for industries that use the product, unlike those in more upstream activities. The logical extreme of this argument are the ultimate upstream activities such as machine tools, to which Rosenberg drew attention[16] – activities that may be small in scale in themselves but have the direct or indirect ability to influence the methods of almost the full range of downstream and final-good activities. To some extent the relative neglect of the civil aircraft industry may have come from a similar implicit view.

This brings me more to the heart of Horrocks's paper. Horrocks concentrates on the use of scientists, especially chemists, in the chocolate industry, notably at Cadbury's. As she notes, much of the work carried out may have been routine testing, but some could lead on to innovation proper. The role of the First World War is interesting, parallelling that more obviously in the aircraft industry. In wartime, demand conditions may indeed change abruptly, as occurred for reasons of supply deficiency in Britain at this time. However many wartime influences on employment in industry were transitory, for example that of women; the situation in regard to scientists, at least in the sectors noted here, is implied in the paper as having been somewhat more permanent, although the final diagram of the paper implies a decline at least from the late 1920s.

In this respect, the paper gives the impression that there was very little product innovation at Cadbury's, once the major brands of Dairy Milk chocolate and Bournville Cocoa were established in the early years of the twentieth century. Apart from the reference to treatment of waste products, plus the wartime activities already mentioned, there is little indication in the paper of anything but stagnation in product development, and

16 Cf. N. Rosenberg, Technological change in the machine tool industry, 1840–1920, in: Journal of Economic History, 23, 1963, pp. 414–43.

hence of the role of innovative chemistry. Perhaps this is a characteristic of the chocolate industry, deriving from the primary importance of entrenched brand loyalty. The rival Mars Bar was jokingly used as a standard of value in the City of London in the 1970s and early '80s, allegedly because its product characteristics were the most invariant of any commercial product from the 1930s onwards.

Most of the innovation alluded to in the paper relating to the interwar period, including nearly all of that in the data appendices, appears in fact to be process innovation – mechanization to raise labour productivity and cut labour and fixed capital costs. This brings us back to the upstream/downstream issue, in that the conventional assumption would be that the food-processing industry was primarily a user of process technologies rather than the inventor. Horrocks points out that there were developments in-house as well as those obtained by searching outside for suitable machinery, and it would be illuminating to expand on this, perhaps in the light of the discussion introducing my comments, and of the work of writers like Lundvall and von Hippel on the "locus of innovation"[17]. More immediately, it implies that the balance of R&D in-house (and bought-in) was shifting from the chemists to the engineers, and perhaps the data might permit the author to elaborate on this.

In the light of contemporary concerns, it would also be valuable to develop the theme, which at present is largely restricted to the data appendix, of impact of labour-saving technical change on the workforce. This is of particular interest in a strongly paternalist firm like Cadbury's, a point that has already been made for the pre-war period in a discussion of the socialist-leaning Edward Cadbury by Rowlinson[18]. In similar vein, the firm was noted for its early utilization of female labour on a large scale, assisted by having one of Britain's earliest female directors of a large company in the form of Dorothy Cadbury[19]. Again the relationship with mechanization deserves further treatment – in my diagram this is the labour process component of the organization sphere.

Finally the competitive environment needs to be addressed. Were other chocolate manufacturers developing new products? Were costs at Cad-

---

[17] Cf. B.-A. Lundvall, National Systems of Innovation: Towards a Theory of Innovation and Interactive Learning, London 1992 and E. von Hippel, The dominant role of users in the scientific instrument innovation process, in: Research Policy, 5, 1976.

[18] Cf. M. Rowlinson, The early application of scientific management by Cadbury, in: Business History, 30, 1988, pp. 377–95.

[19] Cf. C. Dellheim, The creation of a company culture: Cadburys, 1861–1931, in: American Historical Review, 92, 1987, pp. 13–44.

bury's falling faster than at their rivals? The finding of fairly similar R&D expenses in other food-processing firms appears in line with other industries in more recent times; an adequate explanation has yet to be produced. Any answer might cast light on why firms of particular sizes congregate in particular industries. The dynamics of why some firms migrate to other activities, like Glaxo, are similarly ill-understood.

To return, therefore, to my opening remarks, I have to conclude that both papers throw some light on the general concerns of technology transfer literature, but both could be considerably extended by adopting a more general framework. The conclusion I draw is that Braun and Edgerton have helped define the arena for military/civilian spillover (or its absence), but the larger questions of spillover of capabilities have yet to be answered. Sally Horrocks indicates the considerable role of QSEs in consumer industries between the wars, but seemingly more for process capabilities than for new products.

Spin-off from British and German Aircraft Technology after the Great War

HANS-JOACHIM BRAUN AND DAVID EDGERTON

In the interwar years the aeroplane was seen as one of the most important products of modern science and technology, and one which would have particularly profound consequences for warfare, societies, economies and polities. Of course, the aeroplane was by no means unique in being granted transforming powers by observers and promoters, but the extent and character of the enthusiasm for it had many special features[1]. The aircraft industry and aircraft technology also differed in some other important ways from the classical new technologies of the early twentieth century. From small beginnings before 1914 the industry mushroomed in wartime, employing hundreds of thousands of workers in Europe by 1918. More than 200,000 aeroplanes were made in the First World War; total production in the prewar years amounted to no more than a few thousand. Aircraft were primarily a military technology, and after the war production fell back drastically such that employment in Europe fell to tens of thousands of workers, with individual firms counting their employees in the hundreds. To the extent that a civil aircraft industry developed after the war it was a nationally-organised and subsidised technology. Aircraft were a national technology by virtue of the fact that they were a technology of power. Unlike the cases of the electrical and chemical industries then, the aircraft industry saw no steady growth; the producing firms tended to be small and nationally, rather than internationally organised.

[1] Cf. D. Edgerton, England and the Aeroplane: An Essay on a Militant and Technological Nation, London 1991; G. L. Mosse, War and the Appropriation of Nature, in: V. R. Berghahn/M. Kitchen (eds.), Germany in the Age of Total War, London 1981, pp. 102–22; R. Wohl, Par la voie des airs: l'entrée de l'aviation dans le monde des lettres françaises, in: Le Mouvement Social 145, 1988, pp. 41–64.

Surprisingly, the development of aircraft technology in Europe in the interwar period is little studied, although in recent years there has been an upsurge in interest[2].

The expenditure on innovation in aircraft technology was a very high proportion of total expenditures on innovation. In Britain, the Air Ministry, responsible for both military and civil aviation, was the single largest R&D-spending institution throughout the period, spending more than any other government department or private firm[3]. The British airship programme of 1924–30 alone resulted in expenditures of over £400,000 on R&D alone[4]. This was more than ICI spent on its oil-from-coal programme, perhaps the largest industry-funded research programme of the interwar years[5]. In Britain, as in most countries, the bulk of R&D effort, and production, was devoted to military aviation. The main exception is not, as is often thought, the United States, but rather Germany between 1919 and 1933. Germany was prohibited from making military aircraft under the Treaty of Versailles.

What then was the effect of this commitment to innovation in aeroplane technology on the development of civil technologies and industries in interwar Britain and Germany? In other words, was there any 'spin-off' from aircraft technology to the wider economy? First, a clarification is required: is the civil aircraft industry to be seen as part of the civil economy or not? The answer is probably no. The same firms tended to make both civil and military aircraft and in many respects the differences between civil and military aircraft were not great. Aeroengines, which,

---

[2] Cf. E. Chadeau, L'Industrie Aeronautique en France: de Bleriot a Dassault, Paris 1987; E. Constant, The Turbojet Revolution, Baltimore 1980, Edgerton, England and the Aeroplane, and T. Hashimoto, Theory Experiment, and Design Practice: The Formation of Aeronautical Research, 1909–1930, Johns Hopkins University PhD Dissertation, 1990. See also H. Trischler, Luft- und Raumfahrtforschung in Deutschland 1900–1970. Politische Geschichte einer Wissenschaft, Frankfurt, New York 1992; H.-J. Braun, Militärische und zivile Technik. Ihr Verhältnis in historischer Perspektive, in: Uniforschung. Forschungsmagazin der Universität der Bundeswehr Hamburg, 1, 1991, pp. 58–66; ibid, Flugzeugtechnik 1914 bis 1935. Militärische und zivile Wechselwirkungen, in: Technikgeschichte 59, 1992, pp. 341–52; ibid., Konstruktion, Destruktion und der Ausbau technischer Systeme zwischen 1914 und 1945 (= W. König (ed.), Energiewirtschaft, Automatisierung, Information. Propyläen-Technikgeschichte Vol.5, Berlin 1992, pp. 172–80.

[3] Cf. Edgerton, England and the Aeroplane.

[4] Cf. Sir P. Masefield, To Ride the Storm: The Story of the Airship R 101, London 1982, p. 486.

[5] Cf. D. Edgerton, Science and Technology in British Business History, in: Business History 29, 1987, pp. 84–103.

though often neglected, represented almost half the cost of an aeroplane, were typically first designed for military use and then introduced into civil service with only minor modifications[6]. Thirdly, the pattern of innovation in both civil and military aircraft appears to be very similar. Thus, although, military and civil aircraft differed in form and function, both relied on the same technological base.

The study of spin-off from the aircraft industry is exceedingly difficult. Works on the technical development of aviation do not dwell on it, and neither do works on recipient industries and technologies. Nevertheless a crude assessment is worth making.

1. The Effect of War and the Military

One of the most difficult aspects of the problem of analysing 'spin-off' is the ideological sensitivity of the topic. Both the belief that the military and war retard and distort innovation, and the belief that all innovation, howsoever funded, is progressive, are deeply rooted in the great ideologies of the West, sometimes the same one. Many liberals, and most marxists, in the interwar years tended to regard war and military funding as retarding influences. Thus the liberal historian of aviation at London's Science Museum argued in the 1930s that war distorted aviation away from its true purpose – the realisation of liberal dreams of world communication. It would have been better for aircraft design if no aircraft had been designed during the war[7]. The leftliberal science journalist J. G. Crowther argued in 1935 that an examination of speed and other records showed that aeronautical progress was slow in the war, especially in comparison with total expenditures on aviation[8]. What these analyses failed to do was to look at expenditures on innovative activity in both war and peace, and, especially, to recognise the extent to which peacetime aircraft development was funded by the military. Aircraft manufacturers, by contrast, tended to claim that the Great War had accelerated technical progress[9]. In Britain,

---

[6] Cf. R. Schlaifer/R. D. Heron, The Development of Aircraft Engines and Aviation Fuels, Boston 1950, p. 48.

[7] Cf. D. Edgerton, The Relationship Between Military and Civil Technology: a Historical Perspective, in: P. Gummett/J. Reppy (eds.), The Relations between Defence and Civil Technologies, Dordrecht 1988, pp. 106–14.

[8] Cf. J. G. Crowther, Aviation, in: Sir D. Hall et al. (eds.), The Frustration of Science, London 1935, pp. 30–41.

[9] Cf. E. Heinkel, Stuermisches Leben, Stuttgart 1953.

the claim that war accelerated technical progress tended to be made in combination with the argument that state design retarded progress[10].

The retardation and distortion analysis appeared again after the Second World War. The German marxist Gerhard Wissmann argued that war interrupted progress in aeronautics since capitalism and its child, imperialist war, per se retarded progress[11]. The liberal analysts Miller and Sawers, in their well-known study on the technical development of modern aviation saw technical aviation as a largely peacetime phenomenon in the civil aircraft industry[12]. Despite its title, their book dealt almost exclusively with civil aviation: they were writing Hamlet without the Prince. Indeed, they argued that the American aircraft industry of the early 1930s revolutionised aircraft technology, because it was both more civil-oriented and more competitive than the European industry[13]. More recently, Mary Kaldor has put forward a clear argument that military funding, especially in peacetime, has so distorted technical development that it has led to the creation of increasingly inefficient military technologies[14].

Curiously enough, the idea that technical progress is independent of the character of the funder of development also has clear roots in the progressive traditions of the West. The idea of 'spin-off' is largely a rhetorical device to assert that this is the case. If military funding or large scale civil funding leads to the development of civil technologies which find their place in the wider economy this is taken as proof that the original technology was uncontaminated by its funders[15]. Thus, even a critic of military funding like Crowther made a list of the civil benefits of aviation research:

"The excellent experimental and mathematical research of such leaders as Prandtl have provided knowledge essential to the design of swift ships besides aeroplanes. Pines has shown how the methodology of Prandtl may be applied to the study of the turbulence in the linings and gases in a Bessemer furnace for manufacturing steel, and how this study may pro-

---

[10] Cf. Edgerton, England and the Aeroplane.

[11] Cf. G. Wissmann, Geschichte der Luftfahrt von Ikarus bis zur Gegenwart. Eine Darstellung der Entwicklung des Fluggedankens und der Luftfahrttechnik, Berlin 1960, pp. 337–42 and G. Wissmann, Imperialistischer Krieg und technisch-wissenschaftlicher Fortschritt, in: Jahrbuch fuer Wirtschaftsgeschichte 1962, pp. 145–58.

[12] Cf. R. Miller/D. Sawers, The Technical Development of Modern Aviation, London 1968.

[13] Cf. ibid., pp. 58, 257.

[14] Cf. M. Kaldor, The Baroque Arsenal, London 1982.

[15] Cf. Edgerton, The Relationship Between Military and Civil Technology.

vide data for the improvement of blast furnace design. The exact study of the chemistry and physics of explosions inspired by the desire to improve aero-engines has led to the development of the powerful schools of physical chemistry led by Semenov, Hinshelwood and Bone. The necessity for economy of material and weight in aircraft structures has demanded more exact knowledge of the properties and strength of metals and materials. This has provided the most profound impetus to modern structural engineering. The refined methods in aircraft design are being applied in the structure of bridges and houses"[16].

Then, and since, aeronautical R&D has been seen as pushing at the frontier of knowledge, by analogy with the aeroplane itself, which opened up new regions to communication. Similarly aircraft technology is itself seen as the 'leading edge' – significantly an aeronautical metaphor – of technological progress in general; as a 'technologie de point'. Within aviation, too, one can see this kind of argument. Record breaking aircraft are often seen as leading the technical path in aviation. Thus the Supermarine racing flying boat which won the 1931 Schneider Trophy is seen as a revolutionary aircraft which paved the way for the Spitfire; the Douglas airliners are seen as the precursors of modern civil airliners. This analysis must be treated with caution; an aeroplane designed to break a record may not add much or indeed anything to the stock of technical knowledge, it may simply represent a particular application of such knowledge. It might also take away technical resources from the design of aircraft of practical use.

2. *Evidence of Spin-Off*

2.1. Aero Engines

The pre-compression of air ingested by an engine through supercharging and turbosupercharging greatly increased its power output. Before the First World War Hugo Junkers in Germany and the Sulzer Brothers Company in Switzerland had used supercharging on land-based diesels. But in the case of aero-engines supercharging had a particular usefulness since it permitted engine power to be maintained in the cold, less dense air at higher altitudes[17]. Auguste Rateau, of gas turbine fame, began working

[16] Cf. Crowther, p. 38 f.

[17] Cf. Constant, p. 122; M. Barthel/G. Lingnau, Daimler Benz. Das Unternehmen, Mainz 1986, p. 72.

on turbosuperchargers in 1915 and the French Air Force used some of his designs in the war. Siemens in Germany and Brown Boveri in Switzerland made geared, multistaged superchargers during the war[18]. Paul Daimler began work on turbosuperchargers in 1915, and, by 1918, they were installed in aircraft[19].

Rateau's work inspired British work from 1915, carried out at the Royal Aircraft Factory (from 1918 Establishment), which led to the development of a geared centrifugal blower which went into service in the Armstrong-Siddeley Jaguar in 1926. In the United States, the chairman of the National Advisory Committee for Aeronautics, W. F. Durand, played a decisive role in transferring Rateau's work to that country. General Electric produced a turbocharger in 1918, and, together with the Army Air Corps, continued development through the interwar years, leading to the production of successful turbochargers in the Second World War. General Electric's experience with turbochargers was instrumental in pioneering jet engines in the United States[20]. However, supercharging was not of great significance outside the aero-engine market. To be sure, some interwar racing cars were supercharged, as were a few very large cars, but it was not a technology of general applicability. But it may be that, as in the case of Daimler-Benz in Stuttgart, the availability of superchargers distracted engineers from developing small efficient engines without superchargers[21].

Two other, perhaps more important, examples of spin-off from aero-engines may be mentioned. In the United States engineers at Chrysler applied the knowledge obtained from the development of high-compression aero-engines to car engines[22]. The use of aluminium in aero-engines was extensive – by the First World War about one third of the weight was aluminium[23] – and led the way to its use in car engines[24].

One of the most direct examples of spin-off came in the case of fuels. In the First World War the problem of 'knocking' became important as

---

[18] Cf. Constant, p. 122.

[19] Cf. Barthel/Lingnau, p. 74.

[20] Cf. Constant, pp. 121–5.

[21] Cf. Barthel/Lingnau, p. 105; E. Eckermann, Vom Dampfwagen zum Auto: Motorisierung des Verkehrs, Reinbek near Hamburg 1981, pp. 110–11, and ibid., 100 Jahre Evolution, in: O. von Fersen (ed.), Ein Jahrhundert Automobiltechnik. Personenwagen, Duesseldorf 1986, p. 31 f.

[22] Cf. J. B. Rae, The American Automobile Industry, Boston 1984, p. 65.

[23] Cf. G. D. Smith, From Monopoly to Competition: The Transformation of Alcoa, 1888–1986, Cambridge 1988, p. 128.

[24] Cf. Barthel/Lingnau, p. 74.

higher performance was sought in aero-engines. In 1916 Royal Dutch Shell funded research by H. R. Ricardo into fuel performance in internal combustion engines; research was also done in France. Based on this work remedies were developed during the 1920s, especially in the USA. In 1921 engineers at General Motors discovered the anti-knock properties of tetraethyl lead, a compound which had been investigated for possible use in chemical warfare[25]. Although aviation fuel became increasingly distinct from 'motor spirit', the latter also benefitted by research created by aeronautical problems. In Britain, at least, leaded fuel was used very little in motorcars before the war, but found a much readier market in aviation[26].

But, in considering spin-off from aero-engines to motor car engines, it is important to note that the two types, already distinct by the beginning of the First World War became increasingly so thereafter. In 1914 a typical production engine had a horsepower of 50–100, already much larger than motor car engines; by the end of the war 200 hp was normal, by 1930 500 hp, and 1000 hp by 1940. A Rolls-Royce Merlin engine had a capacity of 27 litres. A second important difference was that rotary and radial engines, widely used in the First World War, and the interwar years respectively, were of a pattern never adopted for car engines.

The aero-engine and car industries differed in some important respects. A number of important aero-engine firms produced cars as well – for example, Napier, Armstrong-Siddeley and Rolls-Royce – but no mass-producing car manufacturer made aero-engines. Furthermore, the airframe and aero-engine firms were usually distinct from one another and, indeed, there were many fewer types of aero-engine in production that there were airframes. The costs of development of new aero-engines were much larger than for airframes.

2.2. Airframe Materials

Most aircraft until the 1930s were fabric covered. The fabric, usually canvas, was treated with 'dope', a solution of cellulose acetate. This stretched, hardened and waterproofed the fabric. In the First World War the British government financed the establishment of a factory to produce this

---

25 Cf. Constant, p. 120.

26 Cf. Air Commodore F. R. Banks, I Kept no Diary: An Autobiography, 2nd ed., Shrewsbury 1983, pp. 76–122.

material, which led to the creation of British Celanese which became the major supplier of acetate rayon fibre in the interwar years[27].

Aluminium was already an important metal before the First World War, finding a variety of uses, among them in the manufacture of aero-engines, and, during the war, in the manufacture of an increasing number of fixtures in aircraft, for example fuel tanks. But in terms of aircraft structures it was not aluminium, but a treated alloy, duralumin, which became a key material. It was used in Zeppelin airships, and in British and other airships too. The first successful metal aircraft, the J1, was designed in Germany by Hugo Junkers and built in 1915. The German military was rather skeptical about the metal aeroplane, but, in 1916, ordered six fighter planes for trial[28]. Junkers first used sheet steel, but, as that aircraft was quite heavy, switched over to duralumnium. Also, Adolf Rohrbach and Claudius Dornier developed all-metal military aeroplanes during the war. After the war, the prohibition on German aircraft production by the Versailles Treaty hindered further development. Rohrbach built the first four-engined metal monoplane in 1920[29]; Claudius Dornier's giant R-seaplane designs developed into successful commercial seaplanes during the interwar period[30]. Its widespread use in aircraft had to wait until the 1930s. In the 1920s it competed with both steel and wood as a structural material. Indeed, the British R 101 airship was built largely from steel, while its competitor, the R 100, was built from the now traditional duralumin. The refinement of high-strength aluminium alloys certainly promoted research on aluminium, at least in the case of the American company Alcoa, but these alloys were themselves used almost exlusively in aviation[31].

2.3. Aerodynamics

The science of aerodynamics was developed by practicing engineers, academic engineers, and, perhaps above all, by mathematicians and physicists. This work was carried out in elite scientific institutions and government laboratories. In Britain, the universities of Cambridge and Imperial Col-

---

27 Cf. D. Coleman, in: J. M. Winter (ed.), War and Economic Development, Cambridge 1975.

28 Cf. R. Blunck/H. Junkers, Der Mensch und das Werk, Berlin 1942, pp. 95–100.

29 Cf. Miller/Sawers, p. 12.

30 Cf. J. H. Morrow, German Air Power in World War I, Lincoln/Nebraska 1982, p. 200.

31 Cf. M. Graham/B. Pruitt, R&D for Industry: A Century of Technical Innovation at Alcoa, Cambridge 1990.

lege, London together with the Royal Aircraft Establishment and the National Physical Laboratory, were the key institutions; in Germany the University of Goettingen and the TH Aachen led the way. But aerodynamic theory had a relatively early impact on aircraft design, a sustained impact had to wait for the 1930s.

The effects outside aircraft are much more difficult to determine. One might have expected fluid mechanics to have influenced steam turbine design, but according to Frank Whittle it had not done so at the major British turbine maker British Thomson-Houston, at least by the mid-1930s[32].

One of the most interesting spin-offs from aviation was the idea of streamlining. Aircraft were only properly streamlined from the very late 1920s, but the idea of streamlining very quickly acquired a wide cultural resonance: even streamlined buildings appeared. But streamlining did find practical use in the design of powerboats and racing cars in the 1930s. In railways, too, the idea caught on as in the famous British Mallard steam engines of the 1930s. In Germany, a former employee of the airship company Schuette-Lanz, Franz Kruckenberg, built a propeller-driven autorail train, which established a new speed record of almost 150 mph for a rail vehicle. The German railway authorities found the 'railway track Zeppelin' unsafe and, in any case, the propeller propulsion was useless, but it laid the ground for the development of the 1930s' 'Flying Hamburger'[33].

2.4. Production Processes

In Germany, mass production of automobile engines after the First World War was a direct result of the construction of thousands of aeroplane engines[34]. Apart from technical change, changes in the production processes after the First World War were directly linked to innovations during the war: standardisation and the concentration on the mass production of a single product, in the German case of the large twinseat aeroplane powered by a one-hundred hp inline engine. Licensed production and the exchange of vital information between the German aircraft firms under

---

[32] Cf. Sir F. Whittle, Jet: The Story of a Pioneer, London 1953.

[33] Cf. G. von Haeseler, Fliegende Eisenbahnen, Phantastereien um 1930, in: Zug der Zeit – Zeit der Zuege, Deutsche Eisenbahnen, 1835–1985 Vol. 2, Berlin 1985, p. 630 f.; Braun, Flugzeugtechnik 1914 bis 1935, p. 345.

[34] Cf. B. Bellon, Mercedes in Peace and War: German Automobile Workers, 1903–1945, New York 1990, p. 14.

the Hindenburg Program after 1916 is also worth noting. The rationalisation movement of the 1920s has some roots in the organisation of production during the First World War.

Certainly, some engines were produced on a very largescale in all the belligerent countries, for example the American Liberty engine, introduced in 1917. In Britain, however, none of the big producers of aero-engines in the war became mass producers of car engines after it. Rolls-Royce aero-engines during the war were not made from interchangeable parts. As far as airframes are concerned, they too were produced in very large numbers, and particular types were produced in their thousands. But given that they were made of wood, large scale production had little impact on post-war engineering practice.

The interwar aircraft industry, where in Britain at least the dominant material into the mid-1930s was steel, did not contribute much if anything to large scale production technique, though aero-engine production may have stimulated precision engineering. Certainly in the 1930s the British government enlisted the help of motor car firms in the production of airframes and aero-engines, in order to bring experience of mass production to bear. But the experience was not particularly successful: airframes and engines were much more complex than cars and car engines, and the motor firms undoubtedly had difficulties. Indeed it was in the 1930s that, in aircraft production, the learning effect was first recognised. Indeed, even at the height of large scale production during the war, the learning effect was a major determinant of productivity differences between British factories and between British and American factories[35].

3. *Conclusion*

Although several cases of spin-off from aircraft technology into the wider civil economy were quoted above, the overall spin-off effects were not as ample as one might have thought. In view of the fact that the distinction between the military and the civil sector in the aircraft industry is difficult to make this seems to be surprising. Three reservations have to be made, however. The first is that we know too little about the development of aircraft technology, and the technologies it may have affected at the level of detail required to settle the point. The second is that the time scale we

[35] Cf. E. Mensforth, Airframe Production, in: Aircraft Production 9, 1947, pp. 343–50, 388–95.

have been employing is too short: some spin-off effects from the Great War and interwar period may only have become apparent during and after the Second World War. The third is that it might be the spin-off of people rather than technique as such which may be critical.

Two other questions need raising. The first is the extent to which aircraft technology drew on technological development in other areas. Certainly developments in fuel technology and metallurgy made for other reasons had a great effect on aeronautics, and other examples may readily be cited. The second is the extent to which aircraft development crowded out engineering R&D in particular in the interwar years[36]. To what extent did the large commitment to aeronautical R&D retard technological progress in other areas?

---

[36] Cf. Braun, Militärische und Zivile Technik, and ibid.: Technik und Militaerwesen, Mensch und Technik in der Kulturgeschichte, unpubl. broadcast manuscript, 1983.

Technology and Chocolate: Research in the British Food Industry before 1940

SALLY HORROCKS

"The Oompa-Loompas guided the boat alongside the red door. On the door it said, INVENTING ROOM – PRIVATE – KEEP OUT. Mr. Wonka took a key from his pocket, leaned over the side of the boat, and put the key in the keyhole. "This is the most important room in the entire factory", he said. "All my most secret new inventions are cooking and simmering in here! Old Ficklegruber would give his front teeth to be allowed to be inside for just three minutes! So would Prodnose and Slugworth and all the other rotten chocolate makers!"[1]

Roald Dahl's eccentric chocolate entrepreneur, Willy Wonka recognised clearly the importance of technical innovation to his business. A similar clarity of vision has not prevailed among historians, who have rarely considered the food industry as one in which scientific and technical developments played a significant role. Instead attention has been focussed upon those industries whose scientific credentials have seemed more obviously apparent, such as chemicals and electrical engineering. The British food industry in particular has been almost completely ignored, reflecting the more general shortage of detailed studies which examine the role of science and technology in British industry[2]. Existing studies have also tended

[1] R. Dahl, Charlie and the Chocolate Factory, London 1973, pp. 81.

[2] On British industrial research see M. Sanderson, Research and the firm in British industry, 1919–39, in: Science Studies 2, 1972, pp. 107–51, and D. E. H. Edgerton, Science and technology in British business history, in: Business History 29, 1987 pp. 84 ff., and D. E. H. Edgerton/S. M. Horrocks, British industrial research and development before 1945, Economic History Review, forthcoming 1994. A different perspective can be found in D. C. Mowery/N. Rosenberg, Technology and the pursuit of economic growth, Cambridge/Mass. 1989, pp. 98–119.

to concentrate on the role of the semi-autonomous laboratory which was primarily engaged on research work, without looking in detail at the full extent to which firms actually used science and scientists in their production operations. This has led to the neglect of developments which may have been, "unobtrusive, unannounced, unobserved and uncelebrated"[3], but which nonetheless had a profound impact on industrial productivity levels and upon the nature of the products which we use daily.

This paper seeks to redress some of these imbalances by looking at the British food industry, and placing both research and the use of science made by this sector, in the context of British industrial research more generally. It begins with an overview of British industrial research before 1939, establishing the general patterns to be found and identifying the most significant features of the system of research which developed within industry. This is followed by a consideration of the specific case of the food industry, which is in three parts. The first looks at the ways in which scientists first came to be involved with large-scale food manufacture, including the sort of tasks which they carried out and the transition for outside consultants to in-house laboratories. After this the structure of research and development which emerged in the food industry during the interwar period is examined. Finally the activities of scientists employed by the cocoa and chocolate manufacturers, Cadbury, will be discussed, revealing the way in which routine and research work remained intimately related in this industry.

1. British Industrial Science before 1939

British firms in a number of industries employed scientists before the middle of the 19th century, but this did not become commonplace until its final decades. It was during this period that scientific expertise began to establish for itself a secure and permanent place within the firm[4]. Many firms began by using the services to consultants before employing scientists for themselves. The total number of scientists, most of them chemists, employed in British industry is, however, a matter of conjecture. Estimates made for the period before World War I, which give a figure of between 180 and 225 graduates for 1902, have been shown to be

[3] N. Rosenberg, Inside the Black Box, Cambridge/Mass. 1982, pp. 7–8.
[4] Cf. Edgerton/Horrocks, British Industrial R&D.

unreliable[5]. Recent work by Donnelly suggests that the figure was considerably higher[6]. We know very little indeed about the number of active consultants, or about the type of work which was carried out. But by the interwar years industrial chemistry was a recognised profession, and received considerable attention in Pilcher's handbook issued by the Institute of Chemistry. The wide variety of jobs available to the chemist in industry, including analysis, control, and research were outlined, as were the range of industries in which opportunities were available. This included industries such as leather, glass and agriculture as well as the chemical industry itself. Pilcher indicated that, "in practically all productive industries, as in many other spheres of activity, chemists are now recognised as a part of the essential organisation," and held out the hope that chemists would increasingly move into senior management positions within the firm[7]. The Industrial Chemist put the number of chemists in industry in 1933 as between ten and twelve thousand, of which six thousand were members of the Institute of Chemistry[8]. This suggests that the total number of scientists was even larger. There are also difficulties in determining the level of qualifications held by chemists, many of whom were not graduates, but had passed the examinations of the Institute of Chemistry. The role played by scientists differed considerably between industries and between companies, but the majority were engaged in testing and control work, and did little research. What is clear, however, is that the vast majority of British scientists were employed in industry, not universities or government research establishments[9].

Not only has the question of research in industry received considerable more attention from historians than has routine work, but it is an area for which we have much more evidence of employment levels. By the late 19th century several British firms had established research laboratories, and the number continued to expand up to World War I. The attention which was focussed on industrial research during and after the war gave an impetus to many more firms to grant research separate recognition apart from the routine scientific activities which they had been supporting for some time. Some firms established research laboratories isolated from

---

5 Cf. Edgerton, Science and technology, p. 103.

6 Cf. T. Donnelly, Industrial recruitment of chemists from English universities: a revaluation of its importance, in: British Journal for the History of Science 24, 1991, pp. 3–20.

7 Pilcher, The profession of chemistry, Institute of Chemistry, London 1919, pp. 66–81.

8 Cf. Industrial Chemist 8, 1933, p. 37.

9 Cf. Edgerton/Horrocks, British Industrial R&D.

production, with separate staff and buildings, but this was relatively rare. In most firms research remained close to production, both geographically and in the orientation of the work involved. The interwar years saw a considerable expansion of industrial research both through the foundation of new laboratories and the enlargement of existing ones[10]. Most research took place in large firms, and it was concentrated in a few sectors of industry, notably chemicals and electrical engineering. This does not mean that it was not important to a wide variety of industries, but rather that few others regarded such high levels of spending to be appropriate to their needs. By 1939 the majority of the country's one hundred largest manufacturing companies were engaged in scientific research of some sort[11].

Detailed information concerning the extent of research in industry and the employment of qualified research staff is to be found in a survey of industrial research and development carried out by the Federation of British Industries (FBI) Industrial Research Committee (IRC) in 1943, which covered selected years during the period 1930–41. The IRC circulated questionnaires to companies asking for details of the number of staff employed primarily to undertake research and development, and the amount of money spent on these activities. Routine testing work was specifically excluded, but no strict definition of research and development was given. Although many firms had difficulties in furnishing replies to this request for information, a large number were able to do so, and their responses, supplemented by other sources, enable us to establish a picture of research work across British manufacturing industry. The survey found 1381 employees engaged on research in 1930 and 2566 in 1935, the vast majority of these would have been chemists[12]. These figures, when compared to those given above for the total employment of scientists, suggest that the majority of qualified scientists employed by British industry were engaged primarily on routine work[13] (see Table 1 and 2).

Alongside the quantitative data on British industrial research a number of its other features are worthy of note. Scientists in all firms worked

---

[10] The Industrial Chemist in particular carried articles about new laboratories when they opened, and provided detailed descriptions and often plans of the layout. In 1928, for example, the laboratories of Gas, Light and Coke Co., Lyons, and the ICI agricultural research station at Jealotts Hill were featured.

[11] Cf. Edgerton/Horrocks, British Industrial R&D.

[12] Cf. FBI, Industry and research, FBI, London 1943, p. 7.

[13] A more comprehensive analysis of this survey is undertaken in Edgerton/Horrocks, British Industrial R&D.

Table 1: Expenditure and employment of qualified technical staff for research and development in the UK, as revealed by FBI surveys

	1930	1935	1938	1945	1950
firms	422	484	566	420	301
expend. £000s	1,736	2,696	5,442	21,815	23,779
firms	384	432	520	–	301
QSEs	1,381	2,566	4,382	7,894	8,560

Sources; FBI, *Research*, p. 7, FBI, *Scientific and technical research*, p. 9, FBI, *Research and development*, pp. 6–7. The across years, the samples and response rates were different, nor do the figures for employment correspond exactly with the expenditure figures.

within a much wider technological community, drawing on the work of others in both the public and private sector. In some branches of industrial science specialist groups were established which brought together those working in industry, universities and government research establishments. As well as contacts within Britain, many firms had links with companies abroad, both formal and informal, which involved the exchange of technical information. Indeed, a number of foreign owned firms, including Kodak, Mond Nickel and Heinz established research laboratories in Britain, which contributed to the combined technological efforts of the firm on an international level. Collaboration between suppliers and producers was also important[14].

2. *Science and the Food Industry, the Early Years*

The initial growth of large-scale food manufacturing relied little on the systematic application of scientific knowledge. Process control was carried out by skilled operators whose experience was vital to the maintenance of quality, new pieces of machinery were usually developed by independent inventors outside the firm, and novel recipes were either bought from outside or resulted from 'experiments' which took place on a random basis[15]. Chemists were first employed by food companies to carry out routine analysis in the wake of legislation to control the quality of the

[14] Cf ibid.

[15] Cf. Rowntrees Archives (RA), York, papers of J. W. Rowntree, Box 1, Arch 354, 372–74.

Table 2: R&D Employment and Expenditure. Ranked by Expenditure in 1938 of Firms Spending over £25,000

Rank R&D 1938	1930		1935		1938		1941		1945	
	Qualif Staff	£000S	Qualif Staff	£000S	Qualif Staff	£000S	Qualif Staff	£000S	Qualif Staff	£000S
1 ICI*		420	464	625	510	1075	680	1190	1396	3500
2 GEC*	125	100	175					400	250	
3 BTH*	73		104						132	300
4 METRO VICK*	73	95	87	200	117				137	160
5 EMI				167	145	84				
6 PILKINGTONS	13	109	20	114	37	120	34	130	63	
7 LUCAS			84	70	103	102	155	197	171	450
8 MAGNESIUM METALS						90		90		
9 JAS WILLIAMSON	1	1	9	23	14	85	9	13		
9 UNILEVER	30	35	57	65	79	85	85	100	111	200
11 MOND NICKEL	29	29	39	41	59	72	73	82	64	90
12 DUNLOP			85	50	100	65	105	70	77	158
13 COURTAULDS	41	40	51	40	79	60	92	60	132	30
14 BRIT ALUMINIUM		32		40	48	56	50	73	54	80
14 UNITED STEEL*	34	19	86	44	113	56	135	70	32	50
16 MORRIS MOTORS					40	53				
17 ANGLO-IRANIAN	50	35	50	5	5	0	100	120		
18 BOOTS*	29	34	29	34	38	45	68	80	50	90
19 EASTERN OIL	22	30	27	35	35	43	47	57		
20 BROWN FIRTH		35		35		35		35	30	50

Rank R&D 1938	1930		1935		1938		1941		1945	
	Qualif Staff	£000S	Qualif Staff	£000S	Qualif Staff	£000S	Qualif Staff	£000S	Qualif Staff	£000S
21 BUSMC	9	18	9	27	16	32	9	13	22	35
21 KODAK	7	7	22	16	31	32	53	61	54	120
21 ROWNTREES	28	31	34	27	40	32	34	26	20	35
24 COLVILLES*	16	15	31	17	42	31	91	48	30	
24 CALICO PRINTERS	36	23	48	29	51	31	51	39		
26 AEC		30		30		30		30	37	90
26 BRADFORD DYERS			25	25	25	30	29	40		
28 MORGAN CRUCIBLE	27	19	46	22	58	29	53	28	38	50
29 METAL BOX			4	4	28	28	19	24	19	32
29 CADBURY/FRY	21	26	21	27	23	28	15	14	20	30
29 LMS RAILWAY	35	13	41	23	61	28	58	30	60	200
32 BLEACHERS ASSOC	16	16	14	17	15	26	11	26		20
33 LYONS	71	28	70	22	84	25	80	24		

Notes: Ranking is by expenditure for 1938, for which the most complete figures are available. This list is not definitive, there are clearly important companies missing, see Table 4. The figures quoted for several companies differ from those given in other published sources, and they are not necessarily comparable between companies or even for different years for the same company.
Companies marked * appear in the AScW List.

Sources: All figures for 1930–41 are from the FBI Archives, MSS 200/F/3/T1/127 c), expect GEC, BTH and Metro-Vick. For BTH and Metro-Vick see Edgerton, Science and Technology, p. 101 notes 1 and 2. For GEC see Clayton and Algar, GEC, p. 413. All figures for 1945 are from the FBI Archives, MSS 200/F/T2/7/4, expect GEC as above.

nation's food supply which was part of the wider campaign to improve public health in the mid-19th century [16]. The Adulteration of Food and Drugs Act, 1872, was far more powerful than its predecessor of 1860, and led to the appointment of a considerable number of public analysts by local authorities to enforce its provisions. Although enforcement was patchy, and fines generally low, some companies found that a reputation for honest trading paid dividends and used public awareness of the problem in their advertising campaigns[17].

The job of chemical analyst soon spread to industry, which initially relied upon the services of independent analysts and chemical consultants. Gradually large companies began to employ their own fulltime analysts, and by World War I they had become common in large food companies[18]. In 1915 a visitor to the Silvertown confectionery works of the Co-operative Wholesale Society felt that, 'without science today no business, co-operative or otherwise, can permanently succeed'[19]. By this date they had established themselves not only in a defensive role with regard to legislation, but had demonstrated how their skills could be used to ensure consistency in the quality of raw materials and finished products.

The work of these chemists involved a detailed knowledge of the processes and raw materials involved in food manufacture, and placed them in an ideal position to suggest modifications and improvements. When wartime food control restricted the availability of imported ingredients, especially sugar, it was those companies with trained chemists which were best able to compensate for the shortages by employing alternative ingredients without compromising the taste of their products, and by extending their supplies of the usual raw materials. Scientific control of production enabled companies to reduce wastage, another means by which levels of production could be maintained[20]. The later years of the war and those immediately following it saw an increasing number of firms employing their own chemists, as well as the recognition of research as a distinct and important activity within the firm. Consultants, however, continued to be important to many firms. Simultaneously there were a number of government initiatives designed to promote the application of science to indus-

---

[16] Cf. A. S. Wohl, Endangered Lives, London 1984, pp. 52–5.

[17] Quaker food manufacturers in particular traded on their honest reputation, cf. T. A. B. Corley, Quaker Enterprise in Biscuits, Huntley Palmer of Reading, 1822–1972, London 1972.

[18] Rowntrees first appointed a chemist in 1896, and Cadbury did the same in 1901.

[19] The Wheatsheaf 19, 1914/15, p. 67.

[20] Cf. Cadbury Archive (CA), Minute of the Research Committee, 1914–18.

try[21]. The enthusiasm with which a number of food companies formally established research laboratories during this period must thus be attributed to a combination of factors which included the prior existance of chemical departments which had proven their worth during the war, the enlarged scale of enterprises which meant that there was sufficient work to occupy chemists employed fulltime within the firm, and the importance of personal contacts between managers and chemists[22]. Among those companies which formally established research during this period were Glaxo, Lyons, Colman, and the CWS. While government actions played a part in these developments, they were by no means the crucial factor, and it is important to acknowledge the continuity which existed with activities within the firms themselves which had already become well established.

3. The Structure of Research and Development in the British Food Industry in the Interwar Years

As an increasing number of food companies formalised research it became apparent that enthusiasm for this type of venture was greater in some sectors of the industry than in others, and that a few firms were particularly heavy investors in scientific expertise. The distribution of research efforts quickly established a pattern it was to hold throughout the interwar period, which was revealed by data obtained by the Federation of British Industries in 1943[23]. It should be emphasised that these figures, although they are the most comprehensive available, are by no means complete. The FBI itself believed that they "considerably understate the true position"[24] (see Table 3).

The most striking feature of the distribution of both expenditure and personnel connected with specialised research and development work in the food industry is their concentration within a small number of compa-

---

[21] It was at this time that the Department of Scientific and Industrial Research was established, see I. Varcoe, Scientists, government and organised research in Great Britain 1914–16, in: Minerva 19, 1981, pp. 433–63.

[22] The importance of personal contacts is clear in many companies, including Cadbury, Rowntrees and Lyons, see S. Horrocks, Consuming Science: Science, Technology and Food in Britain, 1870–1939, unpubl. PhD thesis, University of Manchester 1993, chapters 4 and 5.

[23] The results of this survey are to be found in the FBI Archives, Modern Records Centre, University of Warwick, Mss 200/F/4/5/2.

[24] FBI, Industry and research, p. 7.

Table 3: The Twenty Largest R&D Spenders in the British Food Industry, 1930–41

COMPANY	RANK	EXPENDITURE £000S				EMPLOYMENT QUALIFIED STAFF			
	1935	1930	1935	1941	1941	1930	1935	1938	1941
LEVER BROS/UNILEVER	1	35	65	85	100	30	57	79	85
ROWNTREES	71	30.5	27.3	31.5	26.2	28	34	40	34
CADBURY	29	26	27	28	14	21	21	23	15
METAL BOX			4	28	24		4	28	19
LYONS		28	22	25	24	71	70	84	80
UNITED DAIRIES		21	21	21	21				
GLAXO			2	11	16	7	12	21	64
MANBRE & GARTON		1	2.5	5	4	2	3	5	3
HEINZ		1.9	2.3	4.5	3.2	2	3	4	5
HORLICKS		1.5	1.5	4.5	4.5	3	3	3	3
APLIN & BARRETT		5	5	4	3	4	4	3	3
VIROL		1.5	2.5	3.5	4	1	1	2	2
CHIVERS		3	3	3.4	4.8	4	4	5	5
HUNTLEY & PALMER	54	3	3	3	3	3	3	3	3
RECKITT & COLMAN	44	1.2	2	2.8	3	3	4	5	5
GEO. BASSETT		2	2.5	2.7	2	2	3	3	2
RANK		2	2	2	2	5	5	5	5
WM. CRAWFORD		1	1.3	1.7	2				
VITAMINS LTD			1.5	1.5	2.2		6	6	6
FOSTER CLARKE		0.7	1.2	1.4	1.7	1	2	2	2
MARSH & BAXTER		1.1	1.4	1.3					

There is evidence, to suggest that the following companies spent more than the lowest include above:
CROSSE & BLACKWELL
TATE & LYLE
DISTILLERS
CO-OPERATE WHOLESALE SOCIETY

The remaining companies spent the following totals during the respective years:

TOTAL	2.9	3.7	5.2	8.5	10	11	13	18
NUMBER OF COMPANIES	7	9	11	12	6	8	9	12
OVERALL TOTAL	167.2	203.4	276.1	274.4	197	250	334	359
NUMBER OF COMPANIES	25	30	32	33	22	26	27	30

Ranking in this list is determined by expenditure in 1938
Rank 1935 refers to the place in the list of largest manufacturing firms in 1935, based on number of employees, see L. Johnman, The large manufacturing companies of 1935, in: *Business History* 28, 1986, pp. 226–45.
Source: MRC, FBI Archives, MSS 200 F/3/T1/127 c)

nies. Firms spending in excess of £10,000 employed around three-quarters of qualified research staff. The position in 1930 reflects the enthusiasm for research generated in the years immediately following World War I, when many of the leading laboratories were established. The depression of the early 1930s did not prevent many firms from increasing their commitment to research, although expenditure fell amongst some of the largest firms. The outbreak of hostilities in 1939 had a distinct impact, as the figures for 1941 indicate.

Research efforts were not evenly distributed across the food industry, but like expenditure were concentrated in a few sectors. These were mainly those such as cocoa, chocolate and jam which had begun factory production at an early date, and which included a number of very large firms. Also well represented were firms whose products claimed medicinal properties, including babyfoods and vitamin products. The canning industry, especially Metal Box, increased its research expenditure considerably through this period. This firm is included here because their research efforts were directed towards the extension of the range of foodstuffs available in cans rather than the process of can manufacture itself. This included identifying the best variety of potatoes to be canned, and research into the possibility of canning bread. They also liased closely with customers, and undertook development research for them. Several manufacturers of prepared meat and fish products also conducted research on a small scale.

Expertise was concentrated in a few firms, with the majority of contributions coming from a small number of authors and companies. Some firms carried out research which enabled them to develop a significant knowledge base which related specifically to the products and process with which they were involved, and a high level of expertise in certain areas was to be found in some companies. At Cadbury and Rowntrees chocolate and cocoa were the focus of attention, at Chivers it was vitamin C, especially under the auspices of Mamie Olliver, while mustard was the centre of attention at Colmans. The chief chemist at Crosse & Blackwell's was an expert on the theory of colloids and emulsions, and wrote a standard textbook which went through five English editions and was translated into several languages[25].

The results of this survey show clearly the distribution of research within the food industry, and the importance of the activities of just a

[25] Cf. W. Clayton, Emulsions and their technical treatment, London 1923.

few firms. This mirrors the situation for British industry as a whole, which has been outlined above. The high level of expenditure of several food companies, especially Lyons and Unilever, places them among the biggest spenders in industry, exceeding that of leading companies in sectors of industry which have more frequently been considered to depend on scientific knowledge, such as photography or pharmaceuticals. This suggests the importance of considering research in terms of individual firms as well as sectors of industry.

4. *Cadbury*

The origins of the Chemists Department at Cadbury illustrate well several of the general points which have already been discussed[26]. The apointment of Norman Booth as chief chemist in 1901 was a clear response to a recognition of the uses to which such skills could be put, which the success of the first research project with which he was involved, the development of a milk chocolate, set the pattern for future research projects. While routine and research work continued to be carried out in conjunction with oneanother, research was gradually accorded a separate status, with the establishment of a research section within the Chemists Department in 1911, and of a Research Committee to coordinate projects carried out in conjunction with the Engineers Department the following year. A separate research laboratory was envisaged, but in practice this continued to house routine work because of continual pressure on space in all laboratories.

Prior to the appointment of Booth to his post the firm had made use of William Chattaway, the superintendent of the manufacturing department of the Society of Apothecaries as a consulting chemist. Booth was among those who had also worked in the Society's analytical laboratory, where George Cadbury jr had spent a period after studying science at University College. When he returned to the firms, George jr had made some efforts to apply his training to production methods. On his retirement it was said that, at this time science first began to supercede rule-of-thumb in the carrying out of many production operations and he pre-

[26] The general history of the firm is covered in I. Williams, The firm of Cadbury, London 1931, and Industrial Record: a review of the interwar years, Cadbury Bros, Bournville, 1947.

pared the first Bournville recipe book before turning his attention to the development of milk chocolate[27].

The programme of research and development which resulted in the launch of Cadbury's Dairy Milk (CDM) in 1905 established many of the features of research work in the firm up to World War II. It brought together both the Engineering and Chemists Departments, whose close collaboration was crucial to its success. Product and process development were carried out in tandem, ensuring a high degree of manufacturability for the product. Initial efforts by the firm to produce a milk chocolate to compete with the popular Swiss milk chocolates which were widely sold in Britain in the 1890 had met with little success, mainly because they were considered to be of inferior quality. George Cadbury jr began experiments to improve this situation in 1898, and was replaced on the chemical side by Booth after his appointment. Working closely with the engineers under Louis Barrow, Booth was able to use his skills to experiment with modifications in recipes and sources of ingredients, and to ensure consistent standards of manufacture. The successful launch of CDM in 1905 marked a triumph for the technical departments, but they continued to be involved in modifications to its production processes and composition. Throughout the aim was to ensure that the brand, which quickly established itself as a leader in its area, remained consistent in taste, texture and quality. By 1913 sales had exceeded those of imported Swiss milk chocolate, and by 1937 had reached a total of 2 million bars daily[28].

The success of CDM helped the Chemists Department to consolidate its position within the firm, and showed its ability to contribute to the development of new products as well as to act as a guarantor of quality. In the years up to 1914 it continued to extend its role within the factory to include work for almost every department. The formation of the Research Committee in 1912 was intended to ensure that research projects, particularly those connected with the development of new processes, which brought together chemists and engineers, were closely coordinated. Work for this committee, however, formed only one part of the work of the Chemists Department, which in the interwar years came to cover all aspects of the firm's operations, both in the Bournville factory and beyond its confines. By considering the full extent of these activities we can get a

[27] Bournville Works Magazine, 1943, p. 119. The remainder of this section is based on the Cadbury Archives relating to the Research Committee, 1912–1939, and the Annual Reports of the Chemists Department, 1902–1939, unless otherwise indicated.

[28] Bournville Works Magazine, 1937, pp. 255–39.

much clearer sense of how this firm came to incorporate the need for the skills of scientists so comprehensively that they came to be seen as indispensible.

The initial focus of Booth's activities as chief chemist was upon the analysis of raw materials. This quickly spread to finished products and to analysis of goods as they passed through the production process. The range of departments who supplied work to the chemists grew steadily to include most of the firm's operations. Separate laboratories were established at each of the firm's overseas factories from the beginning, and close liason was maintained on technical matters. The staff of the laboratory grew as well, leading to a need to extend the amount of space available, and to divide operations between a general laboratory and one intended mainly for research. By 1933 there were a total of 39 staff in the Chemists Department, of which 18 primarily undertook research work, and 21 concentrated on analysis. This division had been formalised by the appointment of A. W. Knapp to head the research section in 1911, although in practice it was very difficult to draw strict lines segregating routine and research work, with both taking place in the same laboratories and being undertaken by the same members of staff.

In time the services of the laboratories came to cover the full extent of operations, from raw materials through production to distribution, and coming from all parts of the factory, including fuel and water supplies as well as welfare services for employees. As standardised procedures were established quality control began to move out of the laboratory and onto the factory floor, releasing qualified staff to perform more complex investigations. Although analysis continued to form the core of the laboratories' work, its staff came to take on variety of functions, including keeping a record of weather conditions, which was used extensively in the investigation of customer complaints, and the development of expertise in entomology to help tackle storage problems.

The research and development activities of the Chemists Department fell into a number of categories, and were undertaken at the request of departments throughout the firm. In collaboration with the Engineering Department, and under the auspices of the Research Committee a wide variety of projects connected with the mechanisation of production processes were undertaken. The role of the chemists in these projects was often to analyse the samples generated by the testing of new machinery to compare results, and to ensure that a consistent level of quality was maintained. In cases where both new products and processes were involved, the chemists experimented with different combinations of recipe and production process. The Research Committee was also responsible for assessing the suitability of ideas from outside the firm, both those offered directly, and those which looked promising from information derived from sources such as the technical press or visits to other firms. Much of the research which was carried out in collaboration with the engineers came close to the needs of production, and was directed towards specific goals. There was little scope for scientists within the firm to pursue investigations which did not fit into the immediate needs of one or other departments, but in the course of these projects a fundamental understanding of the chemistry of the raw materials used by the firm was established.

The development of this firm specific knowledge base was an important asset for the company, which was able to draw extensively on its own earlier work during later investigations. Knapp carried out a series of investigations which sought to establish a scientific basis for processes which had previously been controlled by rule-of-thumb techniques, leading to changes in the processes themselves and the ways in which they were supervised. He also sought a detailed understanding of the fundamental chemical properties of raw materials, and the ways in which this changed during processing. The often general nature of these investigations, and the lack of commercial sensitivity surrounding their results, enabled Knapp to publish his work in this field extensively[29]. They also proved to be invaluable in the search for uses of the waste products generated by the production process, which made a significant contribution to the firm's profitability.

A concern to avoid waste pervaded management thinking throughout the firm, and extended to all aspects of production. The chemists were a useful asset in this, especially when it came to using raw materials. Knapp's work on cacao shell eventually led to the discovery of a lucrative outlet for the material as a cattle food, high in vitamin D, although earlier attempts to extract a number of chemicals from it had been unsuccessful. Methods were devised to extract cocoa butter from dust and sweepings, and compression techniques to remould cold and broken chocolate pieces of chocolate led to further savings. These efforts extended beyond the factory, and included the production of a machine to straighten paperclips, which was estimated to save a pound a week.

---

[29] A full list of the publications of laboratory staff is given in The Bournville Laboratories, Cadbury Bros, Bournville 1937.

One area of the activities of the Chemists Department which is also worthy of specific attention is its relationship with the Advertising Office. This became increasingly important from the late 1920s on, when it was decided to promote the nutritional properties of chocolate, particularly the vitamin D to be found in milk chocolate, and to do so by targeting doctors as important opinion formers. As well as becoming involved in the production of advertising copy, members of the Chemists Department, especially the biochemist, A. Churchman, kept in close contact with the medical profession. Churchman not only prepared literature for this group, he attended medical exhibitions and gave tours of the factory to local groups of the British Medical Association. The food value of chocolate was stressed, and use was made of advertising campaigns for milk itself to reinforce the message. The Advertising Office even went as far as to suggest that, "as far as our visitors are concernced the Chemists Department is our principal concealed asset"[30]. It was not, however, entirely hidden, since a number of advertisements used the image of the white coated scientist to offer reinforcement to the message of purity and quality which they sought to convey.

Research and routine in scientific activities at Cadbury's were closely linked by their objectives, and remained tied to production needs. Although it is impossible to be precise about the overall financial benefits of the programme, its effects in terms of increased productivity are undeniable, and there is strong evidence that all projects were subjected to rigorous financial analysis before being allowed to proceed. To achieve the efficiency which was constantly sought a wide variety of skills had to be drawn upon, and closely coordinated. In the course of their work several members of staff made contributions to the fundamental understanding of foodstuffs, and published their work, but this was never more than an incidental addition to the main project. Many aspects of the activities described at Cadbury show a considerable degree of similarity with the work carried out in other firms, indicating the economic importance of often very simple scientific techniques. The experience of the food industry suggests the way in which research came to be an important aspect of corporate strategy even when there was no perception of a need for a central R&D laboratory operating in isolation from production, especially where new processes and not new products are a major concern.

---

[30] CA, Advertising Department, Minutes of 23/9/39.

5. Conclusion

During the course of his English journey in 1934 J. B. Priestley visited the Cadburys' factory. One of the things which provoked comment was the laboratory, which caused him to state that "it is part of the fantastic quality of our time that what seem trifling bits of frivolity to most of us are of terrific importance to some people. I saw departments where solemn specialists sit in conference over a bit of cocoanut dipped in chocolate or whatever the trifle may be. Men with learned degrees, men with charts, engineers from all quarters of the world, have to be called in to decide the fate of that bit of chocolaty stuff"[31].

Thirty years before Roald Dahl, even Priestley was able to appreciate the important role which science has played in the British food industry. This is not a role we usually associate this branch of industry with, suggesting that a concentration on scientific activity in a limited range of industries and on programmes funded by the state fails to recognise its contribution to other sectors of the economy in which it also played a prominent role. The need to consider research activity from the perspective of the individual firm is highlightened by its concentration in a small number of companies and sectorys of the industry. The clear benefits which were derived by Cadburys and other firms on the food sector as a result of their investment in scientific and technical research contradicts the image of stagnation with which all of British industry has been burdened, and indicates that such assumptions do not stand up to detailed scrutiny. Further studies of this aspect of British economic development are indeed required before we begin to construct any overall analysis of its role. If George Orwell was right when he listed cut-price chocolate and tinned salmon among the mass-produced consumer goods whose low prices had helped to stave off the revolution[32], then some credit must go to the chemists and engineers increasingly employed by food companies, who not only helped to enhance profits, but had a marked impact on the national diet.

---

31 J. B. Priestley, English Journey, London 1934, p. 94.

32 G. Orwell, The road to wigan pier, reprinted London 1989, p. 83.

III.
The Structure of Enterprise and Innovation

Introduction

FRANÇOIS CARON

Two of the articles included in this book, one written by Gottfried Plumpe and the other by Bernd Dornseifer, lead the French reader to raise several questions about innovation in France:

1. What policies motivated a company's activities in its choice of products and procedures?
2. How did businesses organize research?[1]

A preliminary remark: in Germany, the practice of developing business policy out of innovations and findings within particular sectors of an industry – which in French would be referred to as the "filière technique" (technical advances and their integration) – "does not go back to the period between the two world wars: it is a much older phenomenon." Dornseifer notes that towards the end of the 19th century "dyestuffs companies [had] become capable of synthesizing the entire range of organic chemicals, [that] the decisive tool in implementing this [diversification] policy was Research and Development." This observation applies equally well to electrotechnology. Certain sectors of French industry show evidence of similar activity: Saint Gobain's glass production, Schneider's specialized steel. But these were exceptions. French entrepreneurs' strategies remained much more limited than those of their German counterparts. They were acutely aware of the need for innovation and a good number of them did develop specific programs adapted to their limited means. But they did not master, as did Germans in chemical

[1] We could have posited more general questions, such as the relationship between innovation and business size, or the influence of over-all business structure on innovation. But we have prefered to concentrate on research policy.

and electrotechnology, the entire horizon of technical advances in any given field.

While the period was used by German firms to improve methods existing prior to 1914, several French firms, although they did not entirely copy the German example, developed productive and unusually innovative procedures.

1. *Product Strategies*

A comparative analysis of French and German firms brings out similarities as well as differences: the two countries belong to an identical tradition, that of European technicians. In both business cultures, the engineer has a pre-eminent and dominating role at all levels of responsibility, including decision making. Policy problems are approached from a technical point of view. J. P. Daviet has shown that at Saint Gobain, the head technician has always had a more important role than the general manager; engineers managed the Rhône company as early as 1907. In German industries *Kaufleute* are subordinate to *Techniker* and *Direktoren*. French industries are for the most part dominated by *Polytechniciens*. This heavily influenced business practices in all areas of management: for example, self-financing, commonly preferred in France and Germany, is seen as a natural consequence. But a business culture dominated by technicians does not necessarily result in the exclusion of commercial considerations in business management. In a company such as Schneider in France, the commercial dimension, although subordinate, was an integral part of all business sectors. In the thirties, a general reorganization resulted in the creation of autonomous commercial services in most major French industries.

The resulting product strategies can be expressed in three ways:

- Whatever segment of the process be chosen – intermediary goods (semi-products), equipment, or consumer goods – French and German industries began to listen to the consumer to a much greater extent than is currently admitted.
- Diversification policies responded more often to technological considerations, even when constrained by commercial factors, such as competitors' new products.
- There was a great desire for quality in both Germany and France. In the case of France, where the industry perceived itself as a master craft, the pressure for quality was such that it inhibited diversification.

1.1. The Role of the Consumer

Using the examples of Zeiss, Siemens, and IG Farben, Dornseifer shows that the activities of these German companies were preferentially oriented towards the development, production, and commercialization of equipment. Up until the Second World War, the sector of durable consumer goods comprising, for example, household appliances, remained limited by the weak demand for them in Germany in comparison with the United States. Innovators sought primarily to develop new products in sectors involving low turnover, high mark-up goods, at least in part because industries were able to exploit a combination of low salaries and high technical training.

Most striking among the three cases which he discusses is the attention given to the needs and expectations of the product user: at the center of innovation is "application competence". Sectors given the task of testing new products developed rapidly in the electric and pharmaceutical industries.

The market for mass consumption products was not sufficiently dynamic to make their sale profitable in France or in Germany. But this did not have the same consequence on each country's industrial specialization: French industry concentrated on high quality products within an open-structured system. Larger industries grew principally in the area of intermediary products, at a significantly slower rate than in Germany, especially in equipment.

It would seem, however, that a too segmented analysis of production does not account for its real functioning. Users of equipment or intermediary products served a relay function between the producer of goods and the final consumer whose tastes and needs they knew. This oriented the producing company's research. This is also why it is not possible to separate policies concerning consumer products from those concerned with intermediary products and equipment.

In fact, from the end of the nineteenth century on, research in the larger German chemical and electrical firms seriously took into account consumers' needs. Their overall effort had as its goal the perfection of new products and the improvement of existing ones. The case of tinctorial products is particularly significant here: the first concern was to respond to the clearly perceived demand for inexpensive multi-colored fabrics.

Appearances might, however, give a contrary impression. Dornseifer quotes Carl Bosch as thinking that the production of enamels was "too empirical" to interest real chemists; it would be easy to find similar reactions in France. In fact, for a long time, the larger chemical industries in France and Germany refused to take a direct interest in final products; forward diversification was rare. But this does not imply that they did not take an interest in consumers' tastes and needs through intermediary companies. There were, in fact, three actors in the play: the producer of intermediary goods, their user who manufactured consumer products, and the consumer. One of the strengths of German chemical companies in comparison with their French counterparts was that they learned better and earlier to integrate consumer oriented strategies.

The period between the two world wars seems particularly interesting in France at this point. Very early on, several industries in the area of intermediary goods introduced methods for developing new uses for their products and for adapting them to clients' needs. The best example of this is "Aluminium Français", created in 1911 with the legal status of an anonymous company by French aluminium manufacturers[2]. Its principal function was to research new areas for the use of aluminium. The business managed to tie research and promotional activities closely together. The "Revue d'Aluminium" (The Journal of Aluminium) was created in 1924 as a consequence. The company developed a policy of cooperation with user industries: Citroën for automobiles, Jean Prouvé for construction, as well as most of the aviation industries. Its goal was to adapt aluminium to the needs of each user and to each of their products. The 1924 report indicated that "the technical services of Aluminium Français should be at the constructor's disposal [...] to perfect samples capable of demonstrating to them [and] to their clients what [could] be done" using aluminium. The manufacturer sought simultaneously to invent new consumer goods and to modify his products to better respond to the demands and desires of the consumer, as perceived by the user.

This strategy was in no way exceptional[3]: during the same period, Saint Gobain also used it. Curved glass for windshields was perfected by one of their engineers, Achille Verlay, in cooperation with the main Citroën research laboratory [4]. After the Second World War, the example of Alu-

---

[2] This information was provided by F. Hachez' study for a Diplôme d'Études Approfondies under my direction as part of the research for her doctoral thesis which will be defended in the near future.

[3] Cf. J. P. Daviet, Une multinationale à la française. Paris 1987.

[4] Concerning the role played by users in innovation, cf. E. v. Hippel, The Sources of Innovation. New York 1988.

minium Français became a model for the plastics industry. Such policies are obviously part of product related competition. They pay constant attention to users who, in turn, are in contact with the consumer.

1.2. Diversification

Such strategies for innovation were particularly natural for France in that they continued aspects of the country's industrial history. The best performing French businesses have always concentrated on discovering the latest human desires. They have been outstanding in proposing new products. The case of automobile industry is, from this point of view, exemplary. In the nineteenth century this type of achievement characterized predominantly, but not excusively, small and middle sized firms.

Dornseifer shows that, in the case of German chemical companies, diversification was both the natural consequence of advances in chemical research and of solutions found to the problem of sub-products. The history of Rhône Poulenc between the two world wars does not differ from this model. Founded in 1928, the company grew out of a merger between the Société Poulenc and the Société des Usines du Rhône. It pursued an effective and systematic diversification policy during the period, a policy which still characterizes it. Its directors perfected and promoted a large numer of new products, each one developed according to a process analogous, on a lesser scale, to the one described by Dornseifer. By developing three related industries – textiles, fine chemistry, and pharmaceuticals – Rhône Poulenc was able to compensate for its research limitations, multiplying patent agreements signed with the world's greatest chemists[5]. It is striking to see such similarities between French and German experiences.

1.3. Quality

Another component frustrated French and German efforts to diversify. One of the major traits of traditional business culture is its identification with a product or a group of products whose quality must be, by definition, the best: glass for Saint Gobain, pipes for Pont à Mousson, aluminium for Péchiney. Such an approach assimilates business and craft, the

[5] Cf. P. Cayez, Rhône Poulenc, Paris 1990.

heritage of artisans. Between the two world wars, mass production methods resulted in a conflict: certain engineers saw them as sacrificing quality to reduce costs. Like the development of mass consumption economies, this gives an explanation to the slow development of French mass production methods. It complemented, however, the industries' subordination to user needs and German industries' subordination to equipment goods. The two cultures echo each other here again in their nearly artisanal conception of the engineer's trade, in its faithfully honoring the wishes and needs of the client.

On the whole, French and German business practices were not as different as they are usually said to be, even though they operated on different scales. In both cases, product development was subordinated to research in synergetic technology, limiting the range of possibilities. Commercial strategies were likewise subordinated to technical advances. Some major French producers of middle goods developed methods to promote the use of their goods, based on a constructive dialogue with users. The larger German chemical and electrotechnological firms were equally interested in establishing permanent relations with their clients and took this into account as they oriented their research. On the whole, even though the two were dominated by technicians, both had to obey two imperatives which indirectly subordinated projects to the needs and wishes of the final consumer: first, the development of new techniques never ceased to open new avenues for product diversification; second, the pursuit of high quality products continued the age-old technical and artisanal culture of both countries which the development of mass production techniques did not eliminate.

2. *The Organization of Research*

During the period of the Enlightenment, the relationship between industry and research was institutionalized. In the nineteenth century, this served as the basis for one of the principle orientations in science and technology. In France as in Germany, the world of scholars was never separated from that of business; the emergence of new areas of activity, such as organic chemistry and electricity only accelerated the interpenetration of the two milieux in both countries. This relationship continued between the two world wars. But the appearance of massive "internalized" research during the last third of the nineteenth century in the largest German chemical and electrical firms was truly revolutionary; during the same period, France did not experience anything comparable. The contrast between the two systems from a quantitive point of view remains significant: the research effort furnished by French businesses between the two world wars has no possible comparison with that of their German counterparts. But it is important to look beyond this point of view; research, even if it was not as institutionalised as in Germany, was none the less present in practice in French industry; it was not isolated from production.

2.1. A Disheartening Comparison

The statistics which allow one to measure the research effort furnished by French businesses, when compared with those of Germany are disheartening. In 1926, IG Farben had at its disposal twenty-five major laboratories employing a total of more than 3,700 people. In France the industrial laboratories never employed more than several dozen people. Not one French business at that time committed more than one percent of its total sales to research, while the major German chemical and electrotechnical businesses routinely spent three or more percent. Plumpe shows that IG Farben filed 2,000 patents a year between 1928 and 1930. During the same period, Rhône Pouienc filed around one hundred. In the field of chemistry, from the end of the nineteenth century on, German businesses profited from the dominant positions of several technological monopolies which their research efforts maintained. The French, having started much later, were only able to win marginal positions.

In certain fields, however, such as in glass and aluminium, French businesses did have an undoubted technical advantage. It is therefore not possible to entirely explain the absence of major research laboratories – centralized or decentralized – by pointing to insufficient financial means. The explanation has rather to do with a different approach to the function of research in business.

2.2. Production and Research: Saint Gobain and Péchiney

Saint Gobain was exceptionally innovative in the field of glass technology between the two world wars. Several of their innovations resulted in the perfection of revolutionary new products, for example, security glass and curved glass for windshields in automobiles, fibreglass, and an entire series of specialized glass adapted to particular uses. These successes can be explained in three ways: an age-old tradition of know-how in glass technology; an extensive system of affiliates which allowed them to take advantage of all advances; finally, the essential reason, the cooperation between the business' general management and the factories in which experimental laboratories had been created. The glass factories were in fact both experimental facilities and factories, which is to say, developmental research laboratories where new procedures could be tried, tested and brought to perfection.

In the area of glass, no distinction was really made between production and research. Each engineer, each factory manager had to demonstrate his capacity to improve the production process or product quality. He could take the initiative himself or be directed to do so by the general management. In such a system, the difference between research and production lost all significance. Most engineers were involved in research and production. Competent in a particular aspect of the production process, they continually introduced modifications without having to consult the central laboratory, without such a process becoming hazardous.

But it is difficult to transpose it into chemistry's domain. An embryonic organization was put in place which included both a central laboratory and a factory based laboratory in Aubervilliers working at a semi-industrial level. These two employed around fifty people, each giving most of their time to quality control. However a small department of "Chemical Studies", attached to the Board of Directors, played an increasingly important role, becoming little by little the real initiator of research projects. It informed the general management of technological advances, proposed new ideas, defined project installations, and coordinated research programs. This small group of engineers was, however, more interested in improving procedures than in perfecting new products, their continuing redefinition of research activities was rather more the consequence of a continuing exchange of ideas between the factories and the chemistry study group than the consequence of a formalized procedure. Saint Gobain did not engage in basic research. The business had, however, developed several large scale research projects in cooperation with other businesses. Its principle concern remained that of modernizing the already existing production procedures.

The first years of the 1930's were particularly difficult for Saint Gobain. The second half of the decade could be described as a revolution under the direction of President Hély d'Oissel. With his technical director Léon Guyard, his views concerning chemistry research proved prophetic. He moved the business towards plastics, pharmaceutical products, and pesticides, aided by a special committee for research problems. A new organization was put into place, based on three principles: intensifying relations between the company and its affiliates, developing a cooperative effort between Saint Gobain and the largest chemical companies in the world, especially American, and most importantly, conceiving a new research organization within the business. It was based on a marked increase in the financial and human resources of the laboratories and resulted in the consequent rapid development of their activities, especially those of the central laboratory.

Research at Saint Gobain was organized in relation to the concept of its integration into the production process, or to say it another way, in the absence of any clear distinction between research and production. The case of Achille Verlay, noted before, is a typical example. The success of such a system relied on the high quality of the French engineers' scientific training. The central idea remained that the engineer who knew the production process perfectly and the problems it involved was the best person to define new research and bring it to fruition, rather than engineers working in laboratories separated from the factory. The role of factory laboratories was strictly second to the engineer, whose scientific training enabled him to find that of solutions. The role of technical and chemical research departments was essentially to create a relationship between the exterior scientific world and the center of the business – its factories and their affiliates – by improving and adapting production procedures to users' needs.

Péchiney's experience resembles Saint Gobain's[6]. Shortly after its creation in 1888, the Société Électrométallurgique Française (or SEME) set up a laboratory in Froges (1890) where Héroult worked on improving his process. In 1906, when the Compagnie des Produits Chimiques d'Alais et de Camargue (PCAC) constructed its factory in Saint-Jean de Maurienne, it also built a laboratory composed of control laboratories which in fact, little by little, participated in the improvement of procedures for perfecting alloys. In the field of aluminium research this quickly created a problem of size, not resolved, as in the United States, by the creation of huge laboratories equipped with adequate materials separated from the factory, but rather by "moving research out of the laboratory into the factory"[7].

---

[6] Cf. L. Calas, La recherche au service de la production d'aluminium, in: Histoire technique de la production d'aluminium. Grenoble 1991, pp. 285–307.

[7] Ibid., p. 288.

This was a possible solution only because "research activities were inherent to engineers' functions." The two business cultures resemble each other on this point.

In 1921, SEME and PCAC merged. A central research service was created at that moment out of the tradition of factory laboratories, which was to focus particularly on chemistry. But in 1922, a service for the "Study and Research of Electrometallurgy" was created, which built a laboratory a year later. It started with approximately fifteen employees; by 1950, that number had increased to sixty. The original factory laboratories were not, however, closed. The Physics Research and Testing Service at Saint-Jean de Maurienne was created in 1926 to develop new alloys and study the effect of ageing on aluminium; it was in turn transferred to Chambéry in 1931 to research problems posed continued in Saint-Jean de Maurienne.

The creation of these structures did not at all reflect a transfer of research from the factory to the laboratory. Advances in the field of electrolysis were made by factory engineers: "innovation such as mastering curves (was) largely dominated by those involved in production," noted Calas. "Such an independent attitude was encouraged by the general management which by centralizing all findings, stimulated competition between the factories"[8]. We can conclude with her that "the combination of quality control laboratories, research services, and factory research was productive."

These two examples, Saint Gobain and Péchiney, show that it is possible to speak of a French model of industrial research. The model belonged to a particular technical culture: research activity was integrated into production; most of its engineers changed from one to the other several times during their careers. The engineers' competence, based on their experience in a specific segment of the production process, allowed them to continually improve procedures without having to consult the central laboratory. Each engineer supervised the technical system and improved it on the basis of his scientific knowledge; his training, therefore, made a great difference. The level to which French engineers were trained was in fact very high. In addition, one idea dominated Franch industrial organization: the production engineer was best placed and most competent to orient research and develop the best solutions given that he encountered the system's deficiencies on a daily basis. A similar attitude was to be found in the major French businesses of the era, largely a product of the École Polytechnique and the École Centrale educational training.

---

[8] Ibid., p. 292.

2.3. The European Approach

The difference in scale between expenditures on research in Germany and France makes any comparison between the two systems difficult. If one is willing to consider not only the figures but also the spirit in which the two systems functioned, then several similarities become apparent. Plumpe and Dornseifer both insist on the role played by technical divisions and specialized committees in defining research at IG Farben. Dornseifer describes the research structures of the same business in two ways: on the one hand, the laboratories' organization was based on a "cross fertilization process" between various research groups; on the other hand, that organization was based on dialogue and cooperation between the laboratories' and the factories' engineers and managers. Technical culture dominated developmental research activities. Interaction in the German system between research centres and production workshops, which allowed managers and engineers to influence research, was similar in spirit to the French system's factory laboratory.

Is it certain that the practice of directly tying research to the factory, to production is archaic? Nothing would seem less sure! In their recent work Technology and the Pursuit of Economic Growth, N. Rosenberg and D. Mowery compared[9] the Japanese industrial research system to its American equivalent. They found that "Japanese organization relies heavily on the localized use of on site information rather than on the hierarchical control of specialized knowledge." They add that the engineers assigned to research "have considerable know-how concerning the manufacturing process." Research is organized as "a coalition of semi-autonomous component units rather than a monolithic entity controlled by a central research department." This would seem to justify French and German practices between the two world wars. The authors suggest that the "Japanese style management" of innovation has its origin in nineteenth and early twentieth century European practices.

If IG Farben did not create a centralized research department, it was because the business did not adopt a multidivisonal structure, whereas, at the same point in time, it was this very structure which made the creation of a research department necessary at Du Pont. The advantages of an American style organization, illustrated here by Du Pont are obvious: once a new product had been perfected by the main laboratory, a remarkably

[9] Cf. D. C. Mowery/N. Rosenberg, Technology and the Pursuit of Economic Growth, Cambridge 1989.

efficient system resulted in the development of further new products because the teams' entire focus was on the development, production, and marketing of products. The risk inherent to the American system is that of separating research from the factories.

France's experience illustrates a point here. The basic concepts behind the organization of research were put into question in the early 1930's. At that time, the plans to restructure research, later to shake the post-war period, were laid. The reforms of the 1940's and '50's were largely borrowed from the American model. One might wonder whether they were not too drastic. In several companies, the changes isolated research and production, creating a situation which contradicted the entire French industrial tradition.

Our reflection brings us to two observations. First of all, between the two world wars, French businesses, whether it be a question of product strategies, the technician's role, or the organization of research and development, introduced practices and procedures which were too quickly condemned in the name of the moderization we now know. It would be more judicious to try to understand that experience in terms of lessons for the future.

In addition, in spite of the contrasts distinguishing the German model – which internalized massive research and directed it towards controlling the entire filiére technique – from its French counterpart – which to a much greater extent maintained nineteenth century practices – a profound similarity brings the two together. In both countries, product strategies were largely dependent on scientific advances in precisely defined areas and diversification strategies were prioritised according to either synergetic or scientific techniques. Again, in the two countries, the larger businesses tended to value pre-production sectors but subordinated their product strategies to the needs of the consumer. Finally, in both countries, the organization of research encouraged cooperation between production and research engineers, a distinction which in the French case, lacks relevance. Neither in Germany nor in France was the factory separated from the laboratory, indeed it functioned as a laboratory upon occasion.

Innovation and the Structure of the IG Farben

GOTTFRIED PLUMPE

The dynamic growth of the chemical industry since the 1950s, and its economic importance as one of the key industrial sectors are based on innovative products and processes developed during the first half of this century. These include high-pressure synthesis, catalytic cracking, synthetic fibres, engineering polymers and chemotherapy, to name just a few of the most important. IG Farben played an important part in innovation in all of these areas[1].

Research and development work carried out under the auspices of IG Farben was thus an important phase in the development of the chemical industry. In its brief existence between 1925 and 1945, IG Farben was one of the most research-oriented companies in the world and an outstanding example of the term "science-based industry". The only other German companies with a comparable research effort were firms in the electrical engineering industry such as Siemens and AEG. IG Farben was thus an exception, and as such generalizations should not be made on the basis of its R&D work. That said, even exceptions are part of the rich tapestry of life and IG Farben, which was one of the world's leading chemical companies, was an extremely important exception.

IG's unique research strength, which lasted well into the 1930s, had its roots in the close links between basic research and industrial technology already manifest in the synthesis of dyestuffs and intermediates in the 1880s and 1890s. The rise of the German dyestuffs industry from the 1880s onwards, its dynamic development and international dominance in the fields of dyestuff synthesis, organic chemistry and pharmaceuticals

---

[1] This paper is based mainly upon the study of the author: Die IG Farbenindustrie AG. Wirtschaft, Technik und Politik 1904–1945, Berlin 1990, particularly pp. 200–432, 471–477, 591–613.

was primarily due to the R&D work carried out by the founders of IG Farben: BASF, Bayer, Hoechst and Agfa. The use of R&D as a strategic method of developing products and markets, and the reinvestment of huge profits not only made these companies uncontested leaders in the world market by 1913, but also gave them a leading edge in the organization of applied industrial research.

The strategic emphasis placed on innovation also played an important part in the merger of these firms to form IG Farbenindustrie AG. The merger was initiated in 1924–25 as part of a rationalization scheme made necessary by altered market conditions and the increase in protectionism after the First World War. The future-oriented outlook of these companies was, however, just as important to the final merger decision as defensive considerations.

In concrete terms, what was at stake was the extremely costly development of high-pressure technology, in particular the synthesis of mineral oil. This was placing an extremely heavy burden on the relatively loose framework of the "Interessengemeinschaft", the agreement governing co-operation between the companies prior to the merger. Underlying this was a long-term strategy whereby the dyestuffs companies were to become diversified companies capable of synthesizing the entire range of organic chemicals. The decisive tool in implementing this policy was R&D, which was therefore the focal point of corporate strategy.

Organizational matters naturally dominated the merger negotiations. Surprisingly, the whole question of R&D and its structural place in the new organization was hardly touched upon. This was certainly not because the top managers considered it of secondary importance. On the contrary, it is a sign that research caused few problems on an organizational level. Over the years, each of the companies which joined together to form IG Farben had developed some sort of a system for organizing and monitoring research, mainly through financing. The projects pursued depended largely on decisions made at Board level, with the Boards each regarding following the progress of specific projects as one of their most important tasks. At the same time, the researchers themselves had some control over the projects developed, as they had to apply directly to the Board for the money to pursue interesting and promising lines of research.

Alongside the relatively unbureaucratic and uncodified way in which such decisions were taken, organized research centred round the companies' large scientific and technical laboratories. In the 1880s, all the big dyestuffs companies had built central research facilities, initially to promote the synthesis of dyestuffs and dyestuff intermediates. Over the years these facilities were systematically extended to incorporate new processes and products such as fertilizers, pharmaceuticals, photographic chemicals, fibres and crop protection agents. When the companies merged to form the IG Farben in 1926 they had a total of 25 large laboratories employing over 3,700 people.

Table 1: Research laboratories at IG Farben in 1926

Field	No.	Employees	Incl. chemists	Costs
Dyestuffs	7	1341	384	5.5
Dyeing	6	1469	291	3.7
Nitrogen	1	279	100	1.3
Inorg. chem.	5	284	90	1.3
Pharmaceuticals	4	213	64	0.8
Fibres/photo	2	113	29	0.3
Total	25	3699	958	12.9

In the mid -1920s, research work was still concentrated on dyestuffs and intermediates, even if the above classifications did not always tally exactly with the work being done there. For example, in the nitrogen laboratories in Ludwigshafen/Oppau, R&D work included the synthesis of methanol and mineral oil, while the dyestuffs laboratories included research into the synthesis of organic products, and active ingredients for crop protection products were handled in the pharmaceuticals laboratories.

The task facing IG Farben was to organize research as efficiently as possible, to utilize potential synergies and to avoid expensive duplication of work. In order to do this, a sort of controlling body, a Technical Committee, was set up at Board level. In the early years (up to 1929/30) the Board of Management had over 40 members making it far too large to act jointly on all matters. Various committees were thus set up to make decisions and oversee the most important fields. The entire Board was merely called on to ratify their actions. R&D was one of the fields managed in this way.

All major R&D projects had to be presented to the Technical Committee, which discussed them and supervised the financial angle. Since IG Farben was not divided into divisions along technical lines until 1929, all the strands came together in the central Technical Committee while the large horizontal units in the organization, the Operating Divisions (Betriebsgemeinschaften), remained largely autonomous. These Operating Divisions had been formed by grouping together the production sites belonging to the original companies, and each operating division/works had its own research facilities.

At the start of the Great Depression, this primarily horizontal structure proved inadequate to the task of keeping a tighter technical rein on the company, so in 1929 the organization was divided into three large divisions on the basis of its activities. Division I basically comprised high-pressure hydrogenation, in other words, nitrogen and nitrogen products, the synthesis of minieral oils, etc. Division II consisted of the classic fields of dyestuffs, intermediates, chemicals and pharmaceuticals and Division III encompassed photographic products and fibres. Initially devised purely as controlling organizations, the Divisions rapidly developed into a firm horizontal structure which had a major impact on the running of the company and its subsidiaries.

Each Division was headed by a member of the Board of Management, who also belonged to the Central Committee of the Board, formed from the complete Board in 1931. The Heads of Division thus belonged to the very top of the management structure. The central Technical Committee retained most of its functions, including overall responsibility for R&D, after the creation of the three Divisions, but the day-to-day running of the business quickly passed into the hands of the Divisions. This meant that from the early thirties all R&D work falling within the competence of the individual Divisions was basically organized within these Divisions. As the Heads of Division were also members of the Central Committee, the necessary coordination was assured at the highest level.

Alongside the hierarchical organization of R&D under the Central Committee, the Technical Committee and the Divisions, cooperation among researchers was naturally essential for day-to-day work. In order to facilitate such cooperation, IG Farben developed an extremely flexible and efficient structure comprising specialist committees for each of the main fields of research. In the early 1930s there were over 20 such committees concerned, among other things, with the organizing and supervising R&D (see Table 2).

This network of committees was a vertical structure covering all areas of research within the company. Its importance lay in the fact that the committees brought together the top men in each field from all over the company. The committees were a forum for discussing work in progress,

Table 2: Committees of IG Farben responsible for R&D

Technical Committee for Inorganic Products
Scientific Intermediates Committee
Analytical Committee
Naphtol AS Committee
Committee on Textile Auxiliaries
Solvents Committee
Plastics Committee
Coatings Committee
Azo Dyestuffs Committee
Tripheyl Methane Dyestuffs Committee
Sulphur Dyestuffs Committee
Committee on Alizarin and Vat Dyestuffs
Indigo Committee
Indigosol Committee
Main Pharmaceutical Conference
Central Pharmaceutical Conference
Pharmaceutical Production Committee
Crop Protection Committee
Artificial Silk Committee

passing on experience and talking about personnel and financial considerations. A member of the Board attended these meetings and minutes were taken to ensure that the management was kept in touch with what was going on. The committee structure was also very flexible and could easily be adapted to changing circumstances by including the relevant specialists, or by setting up new committees or sub-committees.

A good example of how this organizational structure worked was the development of rubber research. Because this field used chemicals, especially acetylene derivatives, that were mainly used as solvents in IG, rubber researchers were initially included in the Solvents Committee. In the late 1920s, when the research work was far enough advanced to start planning technical facilities, a new committee, the Rubber Committee was set up. Over the years, sub-committees were established to deal with specific technical and scientific questions arising from the commercial success of synthetic rubber. A similar pattern can be traced in other fields.

Most significant for the way the company was run was the fact that the committees were part of the relevant divisions, reporting to the Technical Committee and were basically directly responsible to the Board of Management. Information channels and decision-making processes rested on a mixture of personal contact between the various levels, and written reports and documents. The most important single factor was personal communication. The chairmen of the main committees were members of the Technical Committee or even members of the Board of Management. Members of the Technical Committee gave talks to staff, right down to heads of laboratory and project managers, to keep them up-to-date with R&D work. This ensured that they themselves had a good overview of both the technical side and the people working for them.

This considerable technical capacity enabled IG to work independently on most projects. An exception to this rule was, however, hydrogenation technology, where it worked with the U. S. company Standard Oil of New Jersey on the development of electric arc processes. For this purpose the two companies set up a joint venture, the Joint American Study Company (JASCO), in 1929. JASCO operated a pilot plant at Baton Rouge, Louisiana.

From the late 1920s onwards IG Farben also operated a comprehensive licensing policy, both as licenser and as licensee. One notable example was the cooperation with Standard Oil on hydrogenation technology. However, IG also had licence agreements with all the major chemical companies in almost every field in which it was involved. The most important licence agreements were with the Aluminium Co. of America and Dow Chemical on light metals, especially magnesium, and with Du Pont on polyamide. Like ICI, IG Farben produced polyamide fibres under licence from Du Pont, and but for the outbreak of the Second World War, which ended the possibility of negotiation, IG would probably have entered into a licence agreement with ICI to produce polyethylene.

Cooperation with external researchers and research institutes traditionally played an important role in developing the innovative potential of the company. It is no exaggeration to say that there was a close network of cooperation between chemical companies and universities, and that this formed the basis for the development of the German chemical industry. Outstanding examples from the years prior to the merger include the synthesis of indigo and ammonia, the development of the first medicines and chemotherapy. Famous names from this period include Adolf von Baeyer, Fritz Haber, Wilhelm Filehne, Paul Ehrlich and Hermann Staudinger.

The basic research carried out by external institutes was essential for industrial R&D. The universities also ensured a supply of qualified personnel both for the research laboratories and for the production plants. A close and trusting relationship between the company and universities was thus an integral part of IG Farben's research strategy. The importance accorded to this is reflected by the fact that the Board of Management itself took charge of maintaining these contacts. The founding fathers of IG, men like Carl Duisberg, Carl Bosch and Arthur von Weinberg, regarded themselves at least partially as scientists and undertook this task personally. In 1931, when the Central Committee was set up, it was agreed unanimously that one of its nine members should have special responsibility for these research contacts.

IG Farben has often been accused of having no uniform R&D strategy, unlike the Anglo-American corporations. I do not believe that this is true. IG management most definitely did have a basic objective – we need only think of its declared aim of becoming a diversified company capable of synthesizing all organic products. Given the size of the company and above all the diversity of its activities this objective could never manifest itself in the form of a single "grand design". The most important factor behind the company's R&D policy was the fact that all members of the management team shared the basic conviction that R&D was essential for the company's future and that there was general agreement on the basic aims.

In the relatively prosperous years following the merger, when the company was earning high profits, this meant that almost every project received authorization and financial support. Costs expanded accordingly, with research expenses reaching ten per cent of sales in the 1920s, the highest level in the company's short history.

Table 3: R&D expenses at IG Farben 1926–44

Period	Amount RM million	Average per annum	Share of sales per cent
1926–29	529	132	10.3
1930–33	258	65	7.1
1934–38	354	89	6.1
1939–44	622	104	4.1

The 1926–29 figure for research expenditure as a percentage of sales (10.3 per cent) is high both historically and by contemporary international standards compared with the likes of Du Pont and ICI, who were IG Farben's main foreign competitors. This figure declined during the Depression and with the rapid expansion of production facilities in the 1930s, sinking to just over four per cent during the war, although the overall research budget was higher. In other words, research became less intense during the life of IG Farben. During the Depression of 1929–33 this was mainly because drastic savings had to be made as sales and earnings plummeted. This savings drive saw the virtual demise of various research projects, including synthetic rubber, and the company was forced to reduce its research staff. In concrete terms this considerably weakened its research potential. About a third of all university graduates working for IG were employed in research. In 1929/30 there were about 1,100 scientists on the payroll. By 1932 the number had fallen by about 100, partly through redundancies and partly through natural wastage. During the 1930s the number of researchers gradually rose again, reaching a high-point of 1,300 in 1940. During the war it fell back to 1,000. This decline becomes more significant if we compare it with the development of the total workforce. In 1929 research scientists represented just one per cent of the workforce. This rose to 1.5 per cent during the Depression because considerably more labourers and clerical staff lost their jobs than researchers. By 1939 researchers only represented 0.8 per cent of the workforce and by 1944 this had dropped to 0.5 per cent. This data provides another sign of the decline in research activity.

Under the Nazi regime, with its centrally planned economy, the company was only able to make up some of the ground lost during the Depression. During this period research concentrated increasingly on those areas that were the focus of government contracts. In some, for example, synthetic rubber, research work forged ahead. However, the disadvantage of this state support was the loss of independence and concentration on projects beneficial to the goal of a fully self-sufficient economy. For a company like IG Farben, whose development depended to a large extent on its competitiveness in international markets, this was somewhat problematic. However, the relatively short duration of the Nazi regime saved the German chemical industry from permanent damage to its competitiveness. The chemical industry in the former East Germany is a prime example of what long-term subordination to state planning of this type can do to an industry.

In the 1930s there was a major structural change in research in IG Farben. On the one hand, the emphasis shifted from Division I (high-pressure synthesis) to Division II, with work on rubber and plastics gaining in importance. At the same time, there was a sharp drop in the proportion of the research budget allocated to the development of new areas of research. This suggests that the innovation drive was losing momentum, because falling expenditure in these areas meant that fewer new fields of activity could be opened up to the chemical industry. However, this was less a deliberate change of policy than an adaptation to the prevailing conditions, which required industry to concentrate on projects promoting self-sufficiency.

Table 4: Distribution of R&D spending in the IG Farben

Period	Div.1	Div. II	Div. III	New fields*
	– in per cent –			
1927–32	59	36	5	30
1933–38	31	59	10	16
1939–43	33	59	8	12

* included in divisional figures

It is never as easy to quantify the results of R&D and their economic impact as it is to describe organizational structures. It is a well-known fact that the number of patents is only a very rough indicator because it says nothing about the importance of the processes and products they refer to. Nevertheless, the figures give some indication of activity. IG Farben obtained an extremely large number of patents, roughly corresponding to its spending on R&D.

The number of patent applications rose in the 1920s, dropped during the Depression and then increased again through the 1930s. Between 1928 and 1930 IG Farben submitted more than 2,000 patent applications a year in Germany and almost 10,000 worldwide. In the thirties IG Farben had more patents than any other company in the world. At the start of 1939 it held over 5,000 patents in Germany and about 34,000 in other major countries throughout the world, including 5,300 in the USA, about 4,000 in the UK, 3,700 in France, 2,300 in Italy and nearly 900 in Japan. The majority – a good third of the German patents – concerned dyestuffs, while 13 per cent were for organic intermediates and solvents, about 10 per cent each were for inorganic products and medicines, eight per cent for photographic products and fibres and about six per cent for the hydrogenation of mineral oil.

A few details of the innovations made by IG Farben should give a clearer idea of what all this meant. In the 1920s by far the largest project was the synthesis of mineral oil, which was brought to a successful conclusion in 1932. The basic principle comprised catalytic hydrogenation of solid carbon from lignite or hard coal, yielding aviation and motor fuel, lubricants and hydrogenation gas. The high energy requirement meant that these processes were less economical than those based on petrochemical feedstocks, and they were only used in Germany during the Nazi period. However, the R&D work in this field provided a wealth of material that was of wider use to the petrochemical industry and is still important today.

The second most important project was the synthesis of rubber, a project which the companies had worked on prior to the merger and which was taken up again in 1926. The project was largely abandoned during the Depression, but in 1935 came the breakthrough with the development of styrene butadiene rubber. A modified form of this synthetic rubber is the most important commodity rubber produced today. IG Farben was also a pioneer in the development of plastics. Its achievements include polystyrene (1929), the development of polyvinylchloride (PVC) into a usable engineering plastic (1931) and the discovery of polyurethane in 1937. IG also developed a polyamide-based synthetic fibre (Perlon) at around the same time as Du Pont (1935) and in 1941 it produced the first polyacrylonitrile fibre.

Until the discovery of penicillin, IG was also a world leader in the field of chemotherapy. Milestones in this field include the development of antimalarial drugs based on quinolone derivatives, for example, Plasmochin (1924), Atebrin (1930) and, above all, the first anti-bacterial sulphonamide (1935). A pioneering innovation in the field of crop protection, phosphoric acid ester (mid-1930s), also belongs to the IG era.

The development of organic intermediates was extremely important for many areas of the chemical industry from dyestuffs through pharmaceuticals and crop protection agents to films and plastics. The products developed are too numerous to be listed here, but it is no exaggeration to say that the success of IG was largely due to its dominant position in this field and that the organic intermediates were the foundation stone for the development of modern chemical technology. In order to give a complete picture of the breadth of research in the IG Farben, its work on inorganic and metal chemistry and its pioneering role in the development of colour photography should also be mentioned.

Since there are no precises data on the contribution made by new products to the company's sales and earnings, it is not easy to quantify their importance, and above all their economic impact. However, I feel that the company's steady diversification is evidence of the importance of research in its development. Even when the merger was agreed in 1926 the name "Farbenindustrie" (dyestuffs industry) was hardly appropriate for the company as a whole. Nitrogen fertilizers were already generating a larger proportion of sales than dyestuffs, and the establishment of new commercial activities based on successful innovations accelerated this trend, making IG Farben one of the most diversified chemical companies by the 1940s.

Table 5: Diversification of IG Farben share of sales in %

Field	1914	1924	1926	1929	1938	1943
Dyestuffs	63	39	34	28	22	11
Nitrogen	–	33	43	34	15	8
Chemicals	28	19	11	16	21	25
Pharmaceuticals	5	4	5	5	8	9
Photo	3	4	4	7	6	6
Fibres	–	–	–	4	9	7
Plastics	–	–	–	–	2	13
Metals	–	–	1	1	7	8
Fuels	–	–	–	1	8	11
Other	1	1	1	4	1	2
Total	100	100	100	100	100	100

Strategies of Growth and Innovation in the Decentralized Du Pont Company, 1921–1950

DAVID A. HOUNSHELL

1. Introduction

Innovation within the firm is a much-studied but little-understood subject. A critical examination of the literature on innovation and industrial research and development reveals that scholars from various disciplines have tended to approach the subject from single vantage points and have often carried out mono-causal analyses. In trying to understand the nature of innovation within the firm many scholars have failed to understand the complex nature of the firm itself, particularly if the firm is a large, decentralized company in an industry whose scientific and technical based are complex and whose modes of competition are not simple. I would call attention to but two examples, which are among the most frequently cited studics in the literature: Willard F. Mueller's 1962 article in The Rate and Direction of Inventive Activity:, The Origins of the Basic Inventions Underlying Du Pont's Major Product and Process Innovations, 1920–1950[1], and Samuel Hollander's The Sources of Increased Efficiency: A Study of Du Pont Rayon Plants[2].

Based upon his 1955 doctoral dissertation[3], Mueller's article identifies and analyzes twenty-five of the Du Pont Company's "most important product and process innovations" (Mueller's words) made between 1920 and 1950 and which accounted for about 45 percent of Du Pont's sales

[1] Cf. W. F. Mueller, The Origins of the Basic Inventions Underlying Du Pont's Major Product and Process Innovations, 1920–1950, in: The Rate and Direction of Inventive Activity. Economic and Social Factors, Princeton 1962, pp. 323–46.

[2] Cf. S. Hollander, The Sources of Increased Efficiency: A Study of Du Pont Rayon Plants, Cambridge/Mass. 1965.

[3] Cf. W. F. Mueller, Du Pont: A Study in Firm Growth, Ph.D., Vanderbilt University, 1955.

in 1948. Mueller examined each of these innovations and then categorized them as deriving either from Du Pont's research programs or from outside the company. He found that of eighteen new products, Du Pont could be credited with only five and was the co-inventor of another. In his analysis of process improvements, Mueller credited Du Pont with five of the seven innovations on his list. From this analysis, he concluded:

"Du Pont has been more successfull in making product and process improvements than in discovering new products. Except for nylon, [O]r-lon, and neoprene, Du Pont's major product innovations have been based upon technology acquired from others. Next to be considered is the significance of these findings in relation to the frequent statement that Du Pont's bigness has created a perfect environment for inventive activity resulting in important new products and processes and major improvements in existing products and processes. The record during the period of this study does not support such a generalization. Although Du Pont has expanded its research expenditures as it has grown – from slightly under $1 million annually shortly before 1920 to $38 million in 1950 – there has not been a proportional acceleration in the number of important inventions (as defined herein) coming from its laboratories. Nylon still remains its greatest success story. Neoprene, discovered in 1931 [sic], probably has been exceeded only by nylon and [O]rlon; and the latter was an outgrowth of its basic discoveries underlying nylon".

Mueller went on in his conclusions to agree with Richard R. Nelson's 1959 thesis that "though private profit motives may stimulate the firms of private industry to spend an amount on applied research reasonably close to the figure that is socially desirable, it is clear [...] that the social benefits of basic research are not adequately reflected in opportunities for private profit, given our present economic structure"[4].

Hollander's book correlates with Mueller's analysis in some important respects. Looking for the sources of increased efficiency of productivity in Du Pont's manufacture of rayon, Hollander found that minor, inplant, almost-routine improvements added up to significantly greater gains in productivity than did process improvements deriving from Du Pont's rayon research and development laboratories. Hollander's message was clear: Du Pont's industrial R&D did not contribute as much to technological change within the firm as some people had thought. If there were benefits deriving from basic R&D, they were not being appropriated by

---

[4] Mueller, p. 346, quoting R. R. Nelson, The Simple Economics of Basic Scientific Research – A Theoretical Analysis, in: Journal of Political Economy, 1959, pp. 297–306.

the firm but were becoming public property through such routes as conference presentations and publications.

The analyses by Mueller and Hollander (and those of most economists during the 1950s and first half of the 1960s) were strongly colored – one can almost say "tainted" – by the context of the times. First, most scholars continued to be wedded to the view of the relationship between science and technology that Vannevar Bush espoused in his 1945 report to the President of the United States, Science, the Endless Frontier. Bush's report was an advocacy piece, arguing for massive funding of basic research, governed by a scientific elite with essentially no controls or strings attached. His recommendations were predicated on a belief that fundamental scientific research inevitably (but usually after ten to twenty or more years) leads to new technology. This view of the science-technology relationship was only beginning to be critically examined when these works began to appear.

More important – and bound together – were two other concerns in American society at the time. The first was the intense concern shared by some politicians and economists about monopoly power in post-war America. Under the leadership of Thurman Arnold, the late New Deal administration of Franklin Roosevelt had launched a massive offensive against big business by reinvigorating the Antitrust Division of the Justice Department and bringing many, many antitrust cases against big businesses in the United States. This policy was continued by the Truman, Eisenhower, Kennedy, and Johnson administrations. Proponents of the Arnold position believed that big businesses, through their massively supported research and development laboratories, had come to dominate technological development and were using patent laws created during an earlier era of independent inventors to gain and maintain monopoly power. For their part, the leaders of the big businesses under attack argued that modern science and technology had become so expensive and large-scale that only big businesses could afford to perform the research and development work necessary to assure the progress of Western civilization. Mueller's work in particular should be seen in this light, for his study offered a direct challenge to the position of big business leaders, particularly to that of Crawford H. Greenewalt, president of the Du Pont Company, who had argued this position in hearings on monopoly power before the Subcommittee of the House Committee on the Judiciary, in various speeches, and in the Du Pont Company's annual reports. (Du Pont, as other research-intensive firms such as General Electric, Kodak, and AT & T, had found itself facing both civil and criminal antitrust cases launched during the 1940s but continuing into the 1960s.)

The literature on innovation from this period was colored by one other important contextual element: the Cold War. Take away the Cold War, with its resulting massive (and seemingly permanent) shift in funding for research from industrial to government sources, and you cannot understand many of the questions that were being posed in the U. S.-based studies of innovation from this period. Indeed, for those scholars of innovation who were so greatly concerned with the monopoly power of large, research-intensive corporations, the Cold War's shift in funding for basic research seemed to make perfect sense[5].

Together, I would argue, these various contextual factors diverted the attention of scholars away from the more central questions of research, development, and innovation within firms. On the one hand, as David Mowery and Nathan Rosenberg have recently argued, the economists' focus on basic research – conditioned by the factors I have identified above and shaped by their neo-classical economic framework – led to a fixation on the issue of appropriability of research results. This "appropriability regime," as Mowery and Rosenberg call it, has hampered our understanding of the economics of R&D and has resulted in often poor public policy regarding R&D[6]. On the other hand, I would argue that, given the fundamentally different political, economic, competitive, and social contexts of our own time (as compared to the contexts in which the innovation studies of the 1950s and 1960s were written), we should to be wary of the conclusions reached in many of those earlier studies. We should re-examine the past afresh and reach our own conclusions about innovation, R&D, economic growth, and the like. In this paper, I would like to make a start by examining the strategies of growth and innovation within the Du Pont Company in the interwar period.

The central issue is how a firm in a scientifically and technologically based industry builds and maintains what are best called "corporate capabilities" such as that the firm can build and maintain a strong competitive (and, of course, profitable) position in the industry. The history of Du Pont between the wars is highly instructive because it demonstrates that the firm succeeded through no single strategy; it developed no general formula for innovation or growth. The strategies for growth were not

[5] I would include here R. Nelson, The Simple Economics of Basic Scientific Research and K. Arrow's, Economic Welfare and the Allocation of Resources for Invention, in: The Rate and Direction of Inventive Activity, pp. 609–25.

[6] Cf. D. C. Mowery/N. Rosenberg, Technology and the Pursuit of Economic Growth, New York 1989. See, in particular, the Introduction, pp. 3–17.

driven by a single factor, such as basic research. But this is not to say that the company had no strategy. Rather, Du Pont's executives realized that the chemical industry was complex, as was the Du Pont firm itself, and that they would have to be pragmatic at every step of the way.

2. *Du Pont in World War I*

When war broke out in Europe in 1914, the Du Pont Company, an explosives manufacturer, was ideally positioned to reap huge profits from the sale of explosives and military propellants (i. e., smokeless powder) to the European belligerents. Although the firm had suffered a set-back in antitrust court in 1912, the consent degree of 1913 had left it with all of its smokeless powder capacity intact and with about half of its blackpowder and high explosives plants. By the standards of the day, Du Pont was a big business; it ranked in the top twenty U. S. firms.

Yet strategically, the war posed problems for the company's leaders. Since as early as 1908, they had worried about being so dependent on purchases by the government and had, in fact, explored the possibility of diversifying the company, thereby making it something more than an explosives manufacturer. The war compounded the executives' worries. Although once-idle smokeless powder plants now operated at full capacity round-the-clock, and although the company was building capacity as rapidly as possible with virtual guarantees of being able fully to amortize those plants, executives worried about what would happen if peace were to break out – about the dramatic contraction of the company and about remaining an explosives manufacturer.

Technologically and scientifically, Du Pont had been dependent on its cartel agreements with the German Nobel interests and the British Nobel group. This was particularly true for the period before 1902–03, when the company established its first research laboratories for work in high explosives products and processes (the Eastern Laboratory, 1902) and in smokeless powder products and processes, blackpowder, and new products (the Experimental Station, 1903). Other members of the explosives cartel had admired Du Pont for its general organizational capabilities developed after Alfred, Pierre and Coleman Du Pont bought the firm from their elders in 1902 and consolidated and then rationalized about two thirds of the entire U. S. explosives industry, and soon they would recognize the firm's research and development acumen.

As the war continued, as Du Pont built more and more smokeless powder capacity, and as profits lowed in as never before, executives determined that diversification of the firm had become imperative. Initially, they explored products they could make using their capacity and know-how in nitrocellulose technology (smokeless powder is nitrocellulose). But soon they concluded that markets for nitrocellulose products were too limited to employ even a fraction of their company's production capacity. After a thorough study led within the company's Development Department by the former director of the Experimental Station, Fin Sparre, executives approved in early 1917 a new strategy of diversification that was bold in conception and long-term in terms of its execution. The plan called for the creation of a "nucleus of [a] system of new industries" including dyestuffs and allied organic chemicals; vegetable oils, paints, and varnishes; watersoluble chemicals; and industries related to celluloid and to cotton purification. Its architect anticipated the plan would require perhaps a generation to carry it out fully.

3. Diversification

Many corporations have foundered when attempting to diversify. Du Pont might have been one of those failures had it not had the enormous earnings retained during the war. Diversification can be achieved through several means, including acquisition of whole companies, joint ventures with existing companies, technology purchases, and internal generation based on in-house R&D. These strategic moves require knowledge of varying degrees: Which company to buy? How good is the company's technology? What can a company offer in a joint venture that cannot be obtained quickly and easily in-house? What kind of real scale or scope economies or complementarities can be realized? What is the science, technology, manufacturing, and marketing underlying a business, and can one assemble it "from scratch"? Which technology is the best to buy or to develop?

Du Pont's Development Department, headed by Fin Sparre after 1919, oversaw virtually all of the strategic moves, acquisitions, and joint ventures made by the company until his retirement in 1944. Sparre's experience as a chemist, then chief chemist, and then director of the company's Experimental Station, along with his substantial interaction before the war with the R&D units of the world explosives cartel, proved critical. Moreover, the company's Chemical Department – its centralized research division which grew substantially during the war – also played a critical role in the diversification efforts, especially in the company's venture into dyestuffs manufacture in which in-house R&D was heavily relied upon.

Yet not many of Du Pont's earliest whole acquisitions made before 1920 can be considered stunning successes – at least in the short-run. Nor can the venture into dyestuffs be called successful by this date; indeed, some of the company's executives saw this business as a rat hole into which Du Pont was pouring good money. Nevertheless, despite mounting losses in many of these new businesses, made acute by the recession of 1920 and 1921, executives never had second thoughts about the diversification of the company. But they did engage in extensive discussion – and often heated debate – about the organizational effectiveness of the increasingly diversified Du Pont Company.

The debate began as early as 1919 but it reached its final crescendo after the mid-year financial report of 1921 showed the company losing money in all of its businesses except explosives. Diagnoses varied. A younger group of executives finally convinced their more senior colleagues that the problem lay with the functionally organized, centrally controlled company. With increasing diversification, it would no longer do to have a single sales organization responsible for selling explosives to the government, dyestuffs to a host of dyehouses, celluloid plastics to a wide array of jobbers and middlemen, and artificial leather to upholsterers, bookbinders, and automobile manufacturers. The same was true with the manufacturing division and research and development division. In September 1921 the Executive Committee determined to reorganize the top management of the entire firm by creating virtually autonomous departments corresponding to the company's diverse lines of businesses. Each department would be headed by a general manager, who would be responsible for all sales, manufacture, and research in his department. The general manager would be accountable to the Executive Committee for the performance of his department. The Executive Committee would now evaluate the performance of the various departments, serve in an adjudicatory role in disputes between departments, and be responsible for charting and evaluating the larger strategic direction of the company. Thus was born the M-form of the diversified corporation[7].

With the Executive Committee's decision to decentralize the company, an intense skirmish ensued about the decentralization of R&D. The director of the Chemical Department launched a campaign to retain his large

[7] The classic study of Du Pont's decentralization is A. D. Chandler, Strategy and Structure, Cambridge/Mass. 1962.

organization intact but ultimately lost and watched his organization being "dismembered" by the new general managers. The remaining question was whether the company should retain some type of central research organization. The new general managers and their assistants, once they had secured control over their own research and development organizations, allowed the formerly powerful research director to retain a small group of researchers to carry out general research that cut across the company's various lines of businesses. Some thought that this remaining central research organization would mainly serve as a kind of clerical group to index and file the various departmental research divisions' research reports, but under new leadership, it would become a very powerful and well-known research organization. Curiously, the decentralization of 1921 left totally untouched the Engineering Department, the centrally controlled organization that had so brilliantly presided over the construction of the wartime explosives plants and the dyeworks that would become one of the largest chemical plants in the world.

Thus, at the end of 1921, the Du Pont Company had taken the organizational shape that it retained until very recently. The decentralized company could continue to diversify its businesses by merely adding new departments, leaving other parts of the organization virtually untouched. On the staff side of the organization chart, reporting directly to the Executive Committee, were the Development Department, the Central Chemical Department, and the Engineering Department (see Figure 1 and 2) Diversification and innovation at Du Pont would be carried out based on the scientific and technical capabilities developed within these organizational structures. (see Figure 3 and Appendix 1 for information on the evolution of the industrial departments and the acquisitions of the firm.)

4. *Corporate Capabilities in the Decentralized Firm*

Although it is tempting to go down Mueller's list of twenty-five "major innovations" at Du Pont and critique his analysis item by item, space considerations prevent such an approach. Instead, I porpose to take but four examples also used by Mueller and then to reinterpret them in a manner that I believe offers greater explanatory power than that offered by Mueller and which can also guide policy considerations. These examples include organic chemicals, titanium dioxide, nylon, and polyester fibers.

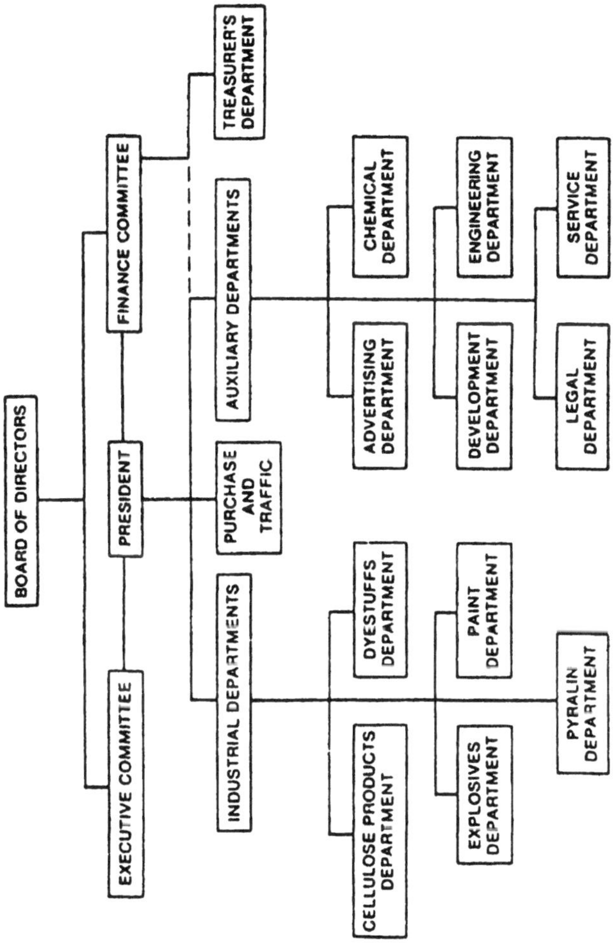

Organization of the Du Pont Company, December 1921. Source: Du Pont Company Organization Charts, Imprints, Hagley Museum and Library.

Figure 1: Du Pont Company Organization 1921

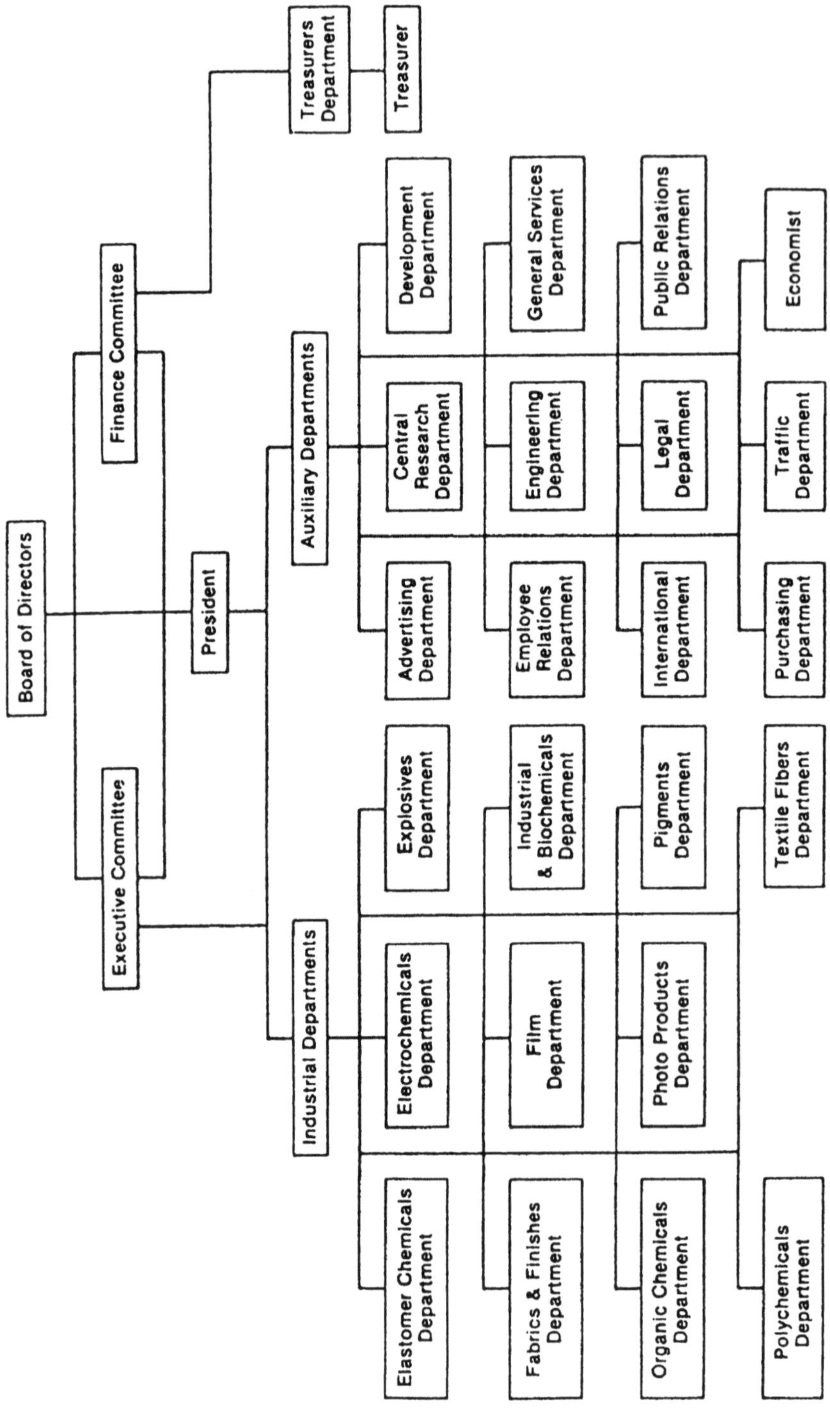

Organization of the Du Pont Company, 1961. Source: Adapted from Du Pont Company Annual Report for 1961.

Figure 2: Du Pont Company Organization 1961

4.1. Organic Chemicals.

Unlike the other diversification efforts made by the company, Du Pont's venture into dyestuffs rested primarily upon its own research and development efforts. The reason was simple. Du Pont could not acquire a synthetic dyestuffs firm because no such industry existed in the United States in 1917; the German dye firms supplied virtually all of the U. S. consumption of dyestuffs. Thus, in a period of a few years, Du Pont tried to carry out the synthesis of products and the development of processes that the German fine chemical industry had developed over a fifty year period. Although Du Pont had access to the knowledge possessed by the comparatively limited British dyestuffs firm of Levenstein, Ltd., and although it also was free to license any of the thousands of German dye patents confiscated by the Alien Property Custodian once the United States entered the war, its leaders learned very quickly about the nature of corporate capabilities.

In a few words, the dyestuffs venture was a trial by fire for Du Pont. The company searched the continent for organic chemists who knew something about dyestuffs synthesis and for businesspersons who knew something about the industry. Between 1917 and 1920 more than two hundred chemists were hired to carry out research and development work at the company's new dyestuffs laboratories, and perhaps as many as twenty-five chemists worked on dye problems out of the Experimental Station.

Progress was slow. Finally in 1920, with the full support of Du Pont's board of directors and under the leadership of the dyestuffs research director, the company took a decisive step by clandestinely recruiting experienced dye chemists from German dye firms to help the firm become a profitable manufacturer of dyestuffs. As the president of the company explained after controversy erupted within Du Pont, the United States, and Germany over the hiring of such chemists, "after two years work in the development of the dye industry we felt sure that many needless experiments could be avoided if there were available in this country men who had practical experience in the dye industry [...] I have no doubt that American chemists in time can solve the same problems which were solved by the German chemists. Neither Germany nor the United States has a monopoly in brains, but there is a grave economic waste, both in money and time, in slowly and laboriously performing over again experiments which have already been made."

The German dye chemists, along with a healthy U. S. tariff on dyestuffs, allowed Du Pont to gain a solid position in the new American industry but did not make it immediately profitable. However, the Dyestuffs Department (later named Organic Chemicals Department) developed significant capabilities in organic synthesis and in the development of organic chemicals manufacturing processes, and these capabilities provided the department with the lever to become one of the most profitable departments in the company. The extensive research work carried out by the department's chemists in its Jackson Laboratory did not lead to any immediate blockbuster products, but it provided the department with the scientific and technical capabilities to perfect a very dangerous process on a commercial scale for the manufacture of tetraethyllead (a process that Standard Oil had devised but had been unable to perfect on a commercial scale), to development processes for the manufacture of Freon refrigerants (a product "discovered" by Thomas Midgely in response to a request by General Motors research director, Charles Kettering), and to develop neoprene synthetic rubber (a discovery of a fundamental research group in Du Pont's central research laboratory). Research in each of these areas yielded still other products and improved processes. For example, the growing expertise of Jackson Laboratory chemists with fluorocarbons (Freon), led to the discovery and development of Teflon polytetrafluoroethylene and other important products based on fluorine chemistry.

Du Pont's top executives drew one very important conclusion from the firm's venture into dyestuffs: It was easier to get into a new business by buying technology or, better yet, a going concern than by researching one's way into it from scratch. Most of them had already reached that conclusion earlier from the company's initial acquisitions of companies in the celluloid plastics field, nitrocellulose coated fabrics, and paints and varnishes. For at least a generation, Du Pont would not try to get into a new business without making an acquisition of some type. Not until the post-World War II era did it attempt this route, and with very few exceptions, these attempts proved to be disastrous. Excellent scientific research alone does not necessarily make for new businesses[8].

4.2. Titanium Dioxide Pigments.

Du Pont had gotten into the pigments business in 1917 when it acquired the Harrison Brothers Paint Company. The later acquisition of the

[8] For a full analysis of Du Pont's entry into the dyestuffs business and the development of its Organic Chemicals Department, see D. A. Hounshell/J. K. Smith, Science and Corporate Strategy: Du Pont R&D, 1902–1980, New York 1988, Chapters 3 and 7.

Grasselli Chemical Company (for quite different strategic reasons) and the Krebs Pigment and Chemical Company allowed Du Pont to become by 1930 the largest U. S. producer of lithopone pigments (a mixture of zinc sulfide and barium sulfate), with a market share of 37 percent. This was no great honor, however, because titanium dioxide pigments (first produced and sold in the United States in 1918) were rapidly penetrating the lithopone market owing to their superior covering power.

The way Du Pont responded to this challenge represents, I believe, the full range of corporate capabilities that allowed the company eventually to dominate this field. Faced with impending destruction of its lithopone assets, Du Pont exercised what Smith and I have termed "development power". The company's Development Department brought into play the scientific and technical units that were associated with lithopone pigments to explore the titanium dioxide situation, while the company's Legal Department assessed the apparently solid patent position. In addition, Du Pont's central research department was employed to explore routes around the patents. Owing to the recently executed Patents and Processes Agreement between Du Pont and Imperial Chemical Industries (ICI), central research was also able to draw upon the research efforts of ICI, whose laboratory at Winnington was then working on titanium dioxide pigments on a semiworks level. The department also retained a chemist who had been in charge of titanium dioxide production in Germany to help clarify aspects of the patent position.

Although all of these efforts came to naught in terms of Du Pont being able to find a clear way around the titanium dioxide patents with a commercially viable process, they allowed the company to make a fully informed and remarkably quick judgement about the necessity of having to form a joint venture with the Commercial Pigments Corporation, which held some important titanium dioxide patents. Through this maneuver, Du Pont was then able to execute a larger patents and processes agreement (later interpreted by the Antitrust Division of the Justice Department as a market-sharing agreement) with the National Lead Company, which also held important patents on titanium dioxide.

The X-ray crystallographic research carried out in 1931 by the central research department during the period of frantically trying to get around the prevailing patents resulted in a major innovation – "rutile" (as opposed to "anatase") titanium dioxide, which possessed twenty percent greater covering power. Du Pont introduced this proprietary product in 1941, and it produced very substantial revenues for the firm.

Finally, Du Pont's position in titanium dioxide pigments, and its continued product and process research in this area, led the company to perfect an entirely new and highly proprietary manufacturing process in the late 1940s. Although tremendous technical difficulties with the new chloride process had to be overcome, Du Pont realized enormous financial benefit from its successful implementation and gained market share to the point that the Justice Department threatened Du Pont with an antitrust case even though Du Pont's domination of titanium dioxide pigments was based solely on it superior manufacturing process[9].

The conclusion to draw from this episode is not that Du Pont research failed and that the company was forced to buy new technology but that Du Pont research provided the means whereby the company could effectively respond to a very serious challenge and could, within a very short period of time, prevail in the new arena in spite of considerable obstacles. This was what corporate capabilities are all about.

4.3. Nylon.

The proponents of the Bush model of the science-technology relationship gained considerable ammunition for their argument that basic research inevitably leads to new technology from Du Pont's invention of nylon (the world's first wholly synthetic fiber), and indeed, they often cited nylon as one of two or three great examples of the basic research/new technology equation (the atomic bomb and the transistor were then other examples). Nylon did indeed emerge from a program of high quality fundamental research in polymer chemistry within the company's central research laboratory, but those who focus on this aspect of the invention miss the most important aspects of the development. Indeed, Du Pont's own executives who charted the company's strategy for the post-World War II period became so enamored with fundamental research that they forgot the reasons why Du Pont succeeded so brilliantly with nylon.

Space considerations prevent a full articulation of nylon's history[10]. But a brief outline is in order. Under the leadership of Charles Stine, the central research department established a small, high caliber fundamental research program focussed on about six areas of chemistry and chemical engineering, including polymers and polymerization. Recruited from an instructor's position at Harvard, Wallace H. Carothers headed Stine's polymer research group. Within a two year period, Carothers published some

[9] For a fuller account of the titanium dioxide episode, see Hounshell/Smith, Chapter 11.

[10] See ibid., Chapters 12 and 13, for the full account.

dozen papers on polymers, including two papers that within a few years were recognized as landmarks, in which he presented a massively documented case that supported the theories of Hermann Staudinger about the nature of polymers. Carothers's papers also provided much of the language and many of the concepts that polymer chemists would use thereafter. And, quite unexpectedly, in a single month in 1930, his group discovered the first synthetic rubber (neoprene – commercialized by Du Pont's Organic Chemicals Department) and the first wholly synthetic fiber (a low-melting, easily hydrolyzed aliphatic polyester). Despite some intense efforts, Carothers and his group were not able to synthesize a practical fiber, but that did not worry him and he went on to explore other basic areas of polymer chemistry. But four years later, when pushed by new management to explore synthetic fibers again, Carothers, guided by his ever-deeper understanding of polymers, outlined a synthesis route to a polyamide. After five weeks of preparing the necessary intermediates, one of Carothers's chemists carried out the polymerization reaction, and the product was a spinnable, high molecular weight, high melting point, highly stable polymer. When spun into fibers, it looked and felt like silk but was far more abrasion resistant. The product was one of many polymers of the polyamide class that was named generically as nylon.

This is where the typical stories of nylon's invention ends. What is lacking is the history of the development of this important fiber. Two vital factors are often overlooked. First, Du Pont had been in the man-made (i. e. cellulose) fibers business for fourteen years when nylon was first synthesized (Interestingly, Du Pont had bought French technology in both viscose fibers and cellulose acetate fibers during the 1920s.). Du Pont's Rayon Department knew the textile fibers business well. With almost one third of the rayon market and a fifth of the cellulose acetate market and multiple manufacturing plants, the department understood the textile market and it knew how to lay out and operate a fibers plant. It possessed capabilities to judge the merits of a fiber, how well a fiber would perform in commercial textile processing machinery, and how consumers might respond to goods made from the fiber. Although the Rayon Department's research organization was seriously lacking in terms of polymer synthesis capabilities (a deficiency fully rectified by the early 1940s)[11],

[11] This is not to say that the department possessed no resarch abilities, for it had assembled an important cellulose chemistry capability and had made a significant improvement in rayon fibers – Cordura tire cord, which proved to be important when synthetic rubber was developed during World War II for tires. Cotton tire cords were not strong enough to handle the higher running temperature of synthetic rubber tires, but Corduar rayon tire cords performed well.

without the knowhow and capabilities of the Rayon Department, the central research unit of Du Pont would have been lost in its attempt to develop Carothers's fiber into a commercial product.

Secondly, without the capabilities that Du Pont's Ammonia Department possessed in high pressure catalytic reactions, the company would have been hard-pressed to develop precesses to manufacture the intermediates from which nylon was to be made. Indeed, one of these two intermediates could not be obtained in the market except in minute quantities and the other was made only in Germany. Du Pont had built its high pressure synthesis capabilities as a result of purchasing a license to the Claude synthetic ammonia process and then trying to straighten out the process in order to make it economically viable. Although the firm did not reap much profit from its synthetic ammonia business per se, profits accrued to the company because of the scientific and technical capabilities built around a strategic acquisition.

Thus, Du Pont's success with nylon rested on much, much more than its achievements in basic research. The capabilities of two industrial departments must share the billing with Carothers and his group.

4.4. Polyester fibers.

Mueller's final product for analysis was Du Pont's Dacron polyester fiber. As he noted, polyester (polyethylene terephthalate) was first synthesized in 1941 by two British chemists, Whinfield and Dixon, who were working for the research association known as Calico Printers Association, Ltd. Unfortunately for Du Pont, Carothers and his followers, based on their experience with the aliphatic polyesters synthesized in the late 1920s and early 1930s, had developed a mindset that polyesters were inherently unstable at temeratures above 100 degrees C and in the presence of water. Whinfield and Dixon shared no such mindset, and after reading Carothers' papers, decided to substitute terephthalic acid for the aliphatic acids used by Carothers and to carry out the polymerization exactly as Carothers himself would have done. The result was highly stable polyester, a versatile polymer that could be meltspun into a fiber of high strength.

Calico Printers did not have the capability to develop this fiber, and in the wartime environment, the British military brought Calico Printers together with ICI in 1943. If any firm in England could develop the fiber, it would be ICI. The wartime emergency had also resulted in a temporary cessation of research information sharing between ICI and Du Pont, which had been an important element in the R&D of both firms since 1929. Consequently, Du Pont did not know of ICI's new development project until one of it consultants, Herman Mark, who was also a consultant to the Army Quartermaster, informed one of the research managers in the Rayon Department in 1944 that ICI was developing a fiber that might be polyacrylonitrile (a fiber that Du Pont was developing as Orlon). Du Pont received two other hints within the next two weeks. One came from a representative of A. D. Little, the Boston consulting firm that was evaluating ICI's fiber for the Quartermaster's Office; the new fiber was melt-spinnable (a fact that eliminated polyacrylonitrile, since that polymer cannot be melted). The second hint again came from Mark; Du Pont could guess the composition of the new fiber merely from the name ICI had chosen for it, "Terrylite". Although the actual name was Terylene, that was enough information for the Rayon Department's Pioneering Research Laboratory to launch a research program to synthesize the ICI fiber. Although at the outset much skepticism remained about the stability of polyester, one of Pioneering's best organic chemists synthesized polyethylene terephthalate, and to everyone's surprise, it possessed ideal fiber properties. During a matter of weeks before a scheduled meeting at which ICI would disclose details about its fiber to Du Pont, the Pioneering Research Laboratory (which had built outstanding capabilities in all aspects of polymer and synthetic fiber research) launched a crash research program and developed data on the fiber that went well beyond the data and knowledge that ICI had gathered during a period of almost two years.

ICI's representatives were stunned to learn that Du Pont had achieved so much with polyester. This produced an outcome that was highly favorable to Du Pont. It resulted in Du Pont's being able to purchase outright (at a very cheap price) the American patent applications of Whinfield and Dixon, thus gaining a monopoly position on the fiber in North America. Such were the R&D capabilities Du Pont was able to marshall.

Moreover, as Hale Charch, director of Pioneering Research, often pointed out during the 1950s, Du Pont's fiber development capabilities were so superior to ICI's that Du Pont was not only the first firm in the world to commercialize polyester, but also that Du Pont's version of polyester (Dacron) was far superior to ICI's Terylene. In fact, until the Justice Department forced Du Pont to abrogate its Patents and Processes Agreement in 1948, Du Pont shared a considerable amount of information with ICI about Dacron's development. Finally, ICI was dependent upon Du Pont's fiber melt-spinning technology that had been perfected after great effort in the nylon development project.

Dacron proved to be one of Du Pont's most important products – second only to nylon in terms of profits. As was the case with titanium dioxide, Du Pont's capabilities in synthetic fiber science and technology allowed the firm to turn a situation that looked highly adverse to the company (polyester did, after all, represent a major threat to nylon) into one of immense gain for the company.

5. Conclusions

In this paper I have sought to argue that by looking at research and development as part of the process by which a firm builds capabilities in science and technology that are employed in a dynamic, competitive environment, we can gain a better understanding of the role that research and development has played within a firm such as Du Pont. Assessing that role must entail something more than measuring inputs and outputs (especially if outputs are constrained by such measures as patents, publications, new products, etc., that economists have traditionally used).

Moreover, I have suggested that the context of the times in which scholars interested in the economics of innovation carried out their work during the 1950s and 60s led them to ask questions about innovation and R&D during the interwar years that were perhaps not really appropriate to this period but more applicable to the Cold War era itself and its patterns of research funding and its conceptions of the science-technology relationship.

The interwar years in the chemical industry are extremely important and interesting. During these years, firms in three nations – Du Pont in the United States, ICI in Britain, and IG Farben in Germany – built chemical empires. None of them did it solely through research and development. Mergers and acquisitions played a central role in the formation and elaboration of these firms. Yet research and development played a critical – and no doubt ever-increasing role – in the process by which these firms became giants and extended their empires.

The Du Pont-ICI Patents and Processes Agreement of 1929 was without question an attempt to preserve and extend the old order that went back to the late nineteenth century British Nobel-German Nobel-Du Pont explosives cartel. But the sharing of research information between these two companies that emerged from the agreement was also extremely important to the innovative capacities of these two firms[12]. To both Du

[12] For more information on this aspect of the agreement, see Hounshell/Smith, Chapter 10.

Pont's and ICI's regret, IG Farben was unwilling to enter into a general patents and processes agreement. But Du Pont and IG Farben did follow each other's technologies.

Du Pont's development into a world-class, diversified chemical firm in the interwar period is important to understand. Acquisition of companies, technologies, patents, knowhow, etc., clearly was the principal means by which the company transformed itself from an explosive manufacturer into a diversified chemical company, such that by 1940 explosives contributed to less than ten percent of sales and even less earnings. But this is not to say that research played no role in this process. In each of these acquisitions, research played a role in deciding which companies and which technologies to buy. Moreover, with most of its acquisitions, Du Pont built research and development organizations around newly acquired technologies. These R&D organizations allowed the company to rationalize the acquired technologies, to advance the manufacturing art, to improve products, and to develop proprietary positions. R&D – and the larger scientific and technological capabilities nurtured and imbedded within the firm – allowed Du Pont to respond to severe threats to its products and markets and, in most cases in the interwar period, to prevail.

In spite of the success of Du Pont's interwar strategy, circumstances forced the company's leadership to develop a new strategy for growth in the post-World War II period. In the context of strong anti-big business sentiments in the U.S. and facing numerous anti-trust indictments, executives concluded that growth by acquisition was out of the question. Ignoring some of the key ingredients in the success of the nylon development and ignoring the sensible advice of the man who created the fundamental research program from which nylon had emerged, these executives plotted a strategy whereby all of the company's industrial department research divisions, not just the central research and central engineering organizations, would engage in extensive fundamental research. Out of the many new laboratories built for this purpose were to flow "new nylons"[13].

This "new nylons" strategy did not succeed in the same way the earlier strategy had. Corporate capabilities, not research and development alone, are what enabled Du Pont to grow and profit in the interwar period. They are also what corporations must still possess in order to compete in world markets of today.

---

[13] For a discussion of the formulation and execution of this new strategy, see ibid., Chapter 16.

Appendix 1: Major Acquisitions of the Du Pont Compay, 1910–33

Year	Acquisition/joint[a] Venture	Products[a]	Price $ million (% Du Pont ownership if less than 100%)	Rationale[b]	Subsequent[c]
1910	Fabrikoid Co.	Artifial leather (nitrocellulose-coated fabrics)	1.2[d]	Outlet for nitrocellulose	Combined with paint business to form Fabrics and Finishes De partment in mid-1920s
1915	Arlington Co.	Celluloid plastics	6.7[e]	Outlet for nitrocellulose	Became Plastics Department in 1921
1916	Fairfield Rubber Co.	Rubber-coated fabrics	0.6[e]	Complement to Fabrikoid	Combined with Fabrikoid
1917	Harrison Brothers and Co.	Paints and chemicals	7.7[e]	Diversification and need for additional acid capacity	Formed basis of Paint Department (1921) and later Fabrics and Finishes Department
	Beton Chemicals	Paints and varnishes	1.1[a]	Diversification into paint business	Becam part of Paint Department
1918	Flint Varnish and Color Works	Paints and varnishes	1.4 (70)[f]	Diversification into paint business	Became part of Paint Department. Minority holding bought in 1921
1920	Chicago Varnish Works	Paints and varnishes	1.0[f]	Diversification into paint business	Became part of Paint Department
	Du Pont Rayon Co. (joint venture)	Rayon fibers	4.0 (60%)[g]	Diversification into highly profitable and growing business	Became Rayon Department in 1936. Du Pont bought minority holding for about $62 million in 1929[h]

1923	Du Pont Cellophane Co. (joint venture)	Cellophane films	1.0(52%)[i]	Diversification, technology similar to rayon's	Became part of Rayon Department in 1936. Minority holding acquired for $12 million in 1929[j]
1924	Lazote, Inc. (joint venture)	Synthetic ammonia	2.3 (63%)[k]	Manufacture of important raw material	Became Ammonia Department in 1929. Du Pont purchased minority holding in 1929
	Du Pont Pahté Film Co. (joint venture)	Photographic film	0.0 (51%)[l]	Diversification (Du Pont had struggled for years to get into business)	Became Photo Products Department after purchase of minority holding in 1941
1925	Viscoloid Co.	Celluloid plastics	2.0 (83%)[m]	Increase market share in celluloid plastics	Became part of Plastics Department. Minority interest acquired in 1928
1926	National Ammonia Co.	Ammonia marketing organization	6.2[n]	Secure marketing organizatin for Du Pont ammonia	Became part of Ammonia Department
1928	Cellulose Acetate Technology	Cellulose acetate fibers	25% of profits[o]	Diversification, establish position in new fiber	Incorporated into Rayon Department. Minority interest acquired in 1936
	Grasseli Chemical Co.	Acids and inorganic chemicals	64.0[p]	Increase market share in heavy chemicals	Became Grasseli Chemicals Department
1929	Krebs Pigment and Chemical Co.	Lithophone white pigment	5.9[q]	Increase market share in pigments	Formed part of Pigments Department
1930	Roessler and Hasslacher Chemical Co.	Electrochemicals	31.6[r]	Provide captive source of metallic sodium used to make other Du Pont products.	Became Electrochemicals Department

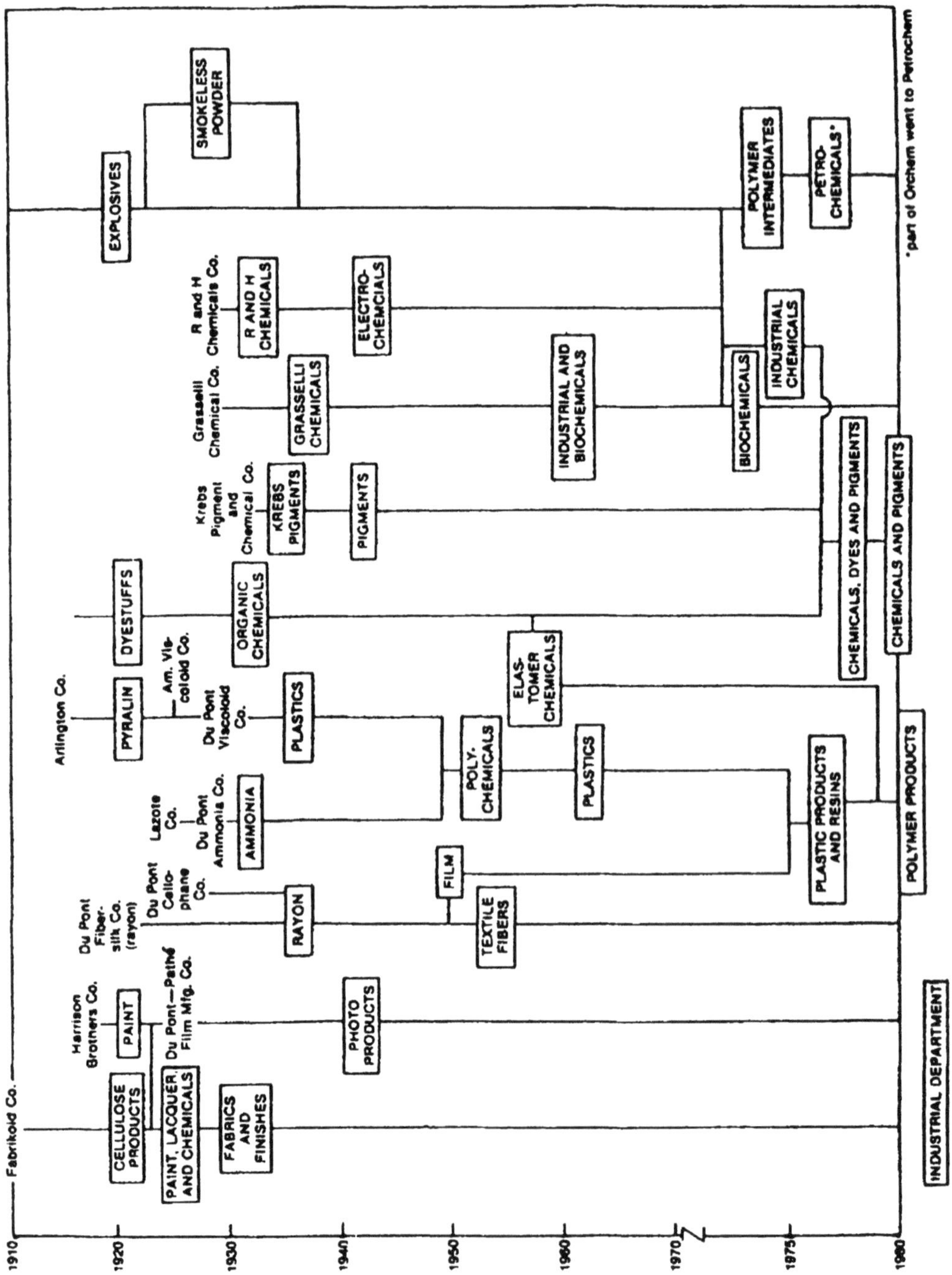

Figure 3: Evolution of the Industrial Departments of the Du Pont Company

Strategy, Technological Capability, and Innovation: German Enterprises in Comparative Perspective

BERND DORNSEIFER

In Germany during the second phase of industrialization two groups of industries attained a particularly high level of international competitiveness: the chemical industry, foremostly dyestuffs, intermediates, and pharmaceuticals, and wide branches of the electrical and non-electrical machinery, equipment and instruments industries[1]. After significant increases in the late 1860s and early 1870s, exports grew rapidly from about 1880 onwards with additional acceleration from the middle of the 1890s. By 1913 the chemical industry exported 35.5% of its output, followed by machinery (26.4%) and electrical engineering (25.5%)[2]. These industries were able to regain their export shares and competitiveness after the war to a surprising extent, despite heavy losses of technology and markets, the rise of powerful foreign competitors and increasing protectionism in international trade. By 1928 the chemical industry fell short of its 1913 share of exports by only 4.2%, and electrical engineering by 6.4%, while the machinery industry had added 2.8%. These industries also recovered most rapidly from the Great Depression. After the heavy devastation of the Second World War they quickly began to display much of their former competitive strength[3]. This line of continuity is one of the most remark-

[1] I would like to thank the participants of the workshop and in particular François Caron for comments on my paper.

[2] These, and subsequent figures on export shares are taken from H. G. Schröter, Die Auslandsinvestitionen der deutschen chemischen Industrie 1870 bis 1930, in: Zeitschrift für Unternehmensgeschichte 35, 1990, p. 2. These industry-wide shares were of course much lower than exports of the leading industry segments and firms which often reached 50 to 70 percent.

[3] Cf. A. D. Chandler, Scale and Scope. The Dynamics of Industrial Capitalism, Cambridge/Mass. 1990, pp. 534–92 on the interwar period, and M. E. Porter, The Competitive Advantage of Nations, New York 1990, pp. 355–68 on the post-World War II era.

able phenomena of modern German economic history. However, the factors which helped to create and sustain that competiveness, the reasons why that very cluster of industries arose in Germany at that time and the problem of how that cluster was shaped by and embedded into the emerging system of international division of labor and national competitive advantage have scarcely been explored.

Despite the fact that we are not in posession of quantitative data which indicate the innovative performance of these industries directly, there can be no doubt – as has often been emphasized – that they were among the most technologically dynamic sectors in German industrialization[4]. Technological capability, it seems, was at the heart of their competitive success. Recent research on the determinants of technological change and international competitiveness has shown that both competitive performance and the capacity to generate technological innovations depend strongly on the particular social, economic, cultural, and political environment within which a firm operates[5]. Furthermore, no country seems to be able to compete and assume technological leadership in every sector. And finally, innovative capability and international competitiveness appear to be shaped in the context of long-term evolutionary processes and structural transformations. These propositions suggest an international com-

---

[4] We do have some data on inventive output of IG Farben in the synthetic materials segment, see C. Freeman, The Economics of Industrial Innovation, 2nd ed. Cambridge/Mass. 1982, p. 58 f., but direct (quantitative) indicators of innovative output are very difficult to obtain. For a recent attempt see Z. J. Acs/D. B. Audretsch, Innovation in Large and Small Firms: An Empirical Analysis, in: The American Economic Review 78, 1988, pp. 678–90. The particular significance of science and technology in German industrialization has frequently been emphasized, see for instance K. Borchardt, Die industrielle Revolution in Deutschland 1750–1914, in: C. M. Cipolla/K. Borchardt (eds.), Europäische Wirtschaftsgeschichte, Vol. 4, Stuttgart 1985, p. 166, p. 182 f.; W. Fischer, The Role of Science and Technology in the Economic Development of Modern Germany, in: P. Lundgreen (ed.), Zum Verhältnis von Wissenschaft und Technik. Erkenntnisziele und Erzeugungsregeln akademischen und technischen Wissens, Bielefeld 1976, p. 179; J. Kocka, The Rise of the Modern Industrial Enterprise in Germany, in: A. D. Chandler/H. Daems (eds.), Managerial Hierarchies. Comparative Studies on the Rise of the Modern Industrial Enterprise, Cambridge/Mass. 1980, p. 96, 99; D. Landes, The Unbound Prometheus. Technological Change and Industrial Development in Western Europe from 1750 to the Present, Cambridge/Mass. 1969, p. 230 and 289.

[5] Cf. N. Rosenberg, Inside the Black Box. Technology and Economics, Cambridge/Mass. 1982, esp. p. 8; W. Lazonick, Organizational Integration in Three Industrial Revolutions, in: A. Heertje/M. Perlman (eds.), Evolving Technology and Market Structure. Studies in Schumpeterian Economics, Ann Arbor 1990, pp. 77–97; Porter, Competitive Advantage of Nations, esp. p. 19 f.

parative approach to the history of industrial innovation and competition, and an exploration of the *longue durée*.

The objective of the present paper is threefold. First, I attempt to show that throughout the period from the late 19th to the middle of the 20th century German enterprises displayed a distinct weakness to innovate in the modern, high-volume consumer goods industries, especially in comparison with American firms. That weakness, I shall argue, did not result from a lack of technological resources but from the retarded growth of domestic consumer goods markets. In the second part I attempt to carve out the specific strategic and structural characteristics shared by the largest enterprises of the internationally most competitive industries. Finally the third part consists of an examination of three competitive companies operating in that cluster of industries: Zeiss (scientific instruments), Siemens (electrical machinery and equipment) and Bayer/IG Farben (dyestuffs, synthetic materials). The purpose of these studies is to determine the prevailing patterns, organizational centres, and sources of innovation, and to shed some light on how the firms' innovative activities were embedded into their competitive and corporate strategies[6].

1. *The Continuing Disadvantage of a Small Domestic Consumer Goods Market*

The key to understanding the specific pattern of innovative strengths and weaknesses of the German economy before World War II is the fact that the overwhelming majority of the country's largest industrial enterprises continued to cluster in the capital goods sector. Alfred D. Chandler's data show that in 1913 only 13 of the largest 100 German industrial corpora-

[6] In the course of my work I have come to the conclusion that innovation should not be viewed as a generic firm strategy. If successful, innovative activities lead either to enhanced diversification or vertical integration up- or downstream (corporate strategies) or to product differentiation or price leadership (competitive strategies). Competitive and corporate strategies are closely intertwined. Vertical integration as well as (interrelated) diversification usually has significant effects on a firm's cost structure or its potential for product differentiation. In most cases innovative activities extending vertical integration are merely instrumental to diversification, product differentiation or price leadership strategies. By and large different competitive strategies put an emphasis on different innovative activities. While price leadership strategies foster process innovations and standardization of product, differentiation strategies direct efforts towards improving the technical features and quality of the product. For these conceptual considerations see: M. E. Porter, Competitive Advantage. Creating and Sustaining Superior Performance, New York 1985, pp. 164–200.

tions – ranked by capital value, mining and gas extraction industries not included – produced and marketed primarily consumer goods for direct use by individuals and households[7]. By 1929 their number had grown to 24, but by 1953 had dropped to 21 (West Germany), as a result of the war.

In the United States, in contrast, as many as 56 of the top 100 group already concentrated on consumer goods in 1917. By 1930 their number had grown to 60, and by 1948 decreased slightly to 58. In Great Britain as well representation of the consumer goods sector was much stronger than in Germany, but at first not as prevelant as in the United States. There, 43 companies concentrated on consumer products in 1919, 60 in 1930, and 59 in 1948. Only in France, it appears, did consumer goods producers remain as rare among the 100 largest industrial enterprises as they did in Germany. There, however, enterprises continued to be generally much smaller in size when compared with the other three nations until after World War II[8].

---

[7] These and the following figures on sectoral clustering have been drawn from Chandler, Scale and Scope, Appendices A.1-A.3 on the United States, B.1-B.3 on Great Britain, and C.1-C.3 on Germany. The following industry groups have been assigned to the consumer goods sector: 20 (Food and kindred products); 21 (Tabacco manufactures); 23 (Apparel and other textile products); 251 (Household furniture); 264 (Misc. converted paper products); 271 (Newspapers), 272 (Periodicals), 273 (Books), 274 (Miscellaneous publishing), 277 (Greeting card publishing), 278 (Blankbooks and book binding); 283 (Drugs), 284 (Soap, cleaners, toilet goods), 285 (Paints and allied products); 291 (Petroleum refining); 301 (Tires and inner tubes), 302 (Rubber and plastics footware); 313 (Boot and shoe cut stock and findings), 314 (Footwear), 315 (Leather gloves and mittens), 316 (Luggage), 317 (Handbags and personal leather goods), 319 (Leather goods); 321 (Flat Glass), 322 (Glass, glassware), 326 (Pottery and related products); 341 (Metal cans, shipping containers), 342 (Cutlery, hand tools, hardware), 343 (Plumbing, heating); 363 (Household appliances), 3636 (Sewing machines), 364 (Electric lighting and wiring equipment), 365 (Radio and TV receiving equipment), 366 (Communication equipment); 3711 (Motor vehicles), 3714 (Motor vehicle parts, accessoires), 375 (Motorcycles, bicycles, parts); 385 (Ophthalmic goods), 386 (Photographic equipment and supplies), 387 (Watches, clocks, watchcases); 391 (Jewelry, silverware, plate ware), 393 (Musical instruments), 394 (Toys, sporting goods), 395 (Pens, pencils, office, art supplies), 396 (Costume jewelry and notions), 399 (Misc. manufacturers). One can of course argue about the assignment of individual groups: numerous groups had both consumer and capital goods segments and some have shifted their main capacity from the capital goods to the consumer goods sector over time. Others, such as communication equipment, had the institutional characteristics of a consumer goods industry in one country and those of a capital goods industry in an other. What matters here, however, is consistency of classification in order to get a rough picture of the major international similarities and differences.

[8] Cf. M. Levy-Leboyer, The Large Corporation in Modern France, in: Chandler/Daems (eds.), Managerial Hierarchies, pp. 122–32. Levy-Leboyer, p. 131, estimates that by 1968 not more than 38 of the 100 largest French companies (mining and extraction included) were clustered in the modern high-volume consumer goods industries.

International differences were less marked with respect to the corporations ranked 101–200 in each country[9]. In Germany, 38 firms of that group concentrated on consumer goods in 1913, 47 in 1929, and 43 in 1953. In the United States, representation was only somewhat stronger, except during the interwar period: 43 in 1917, 43 in 1930, and 51 in 1948. Conversely, in Great Britain, more firms continued to be clustered in the consumer goods sector than in both the United States and Germany: 50 in 1919, 54 in 1930, and 58 in 1948.

The weak representation of consumer goods producers especially among Germany's top 100 enterprises reflects not only the outstanding growth of capital goods suppliers, but also the retarded evolution of high-volume consumer goods markets in that country. The growth of consumer goods markets was delayed by various factors. Throughout the late 19th and early 20th centuries wages were lower than in the United States by a wide margin, lower than in Great Britain, and also probably slightly lower than in France[10]. The impoverishment of the population caused by the First World War and its economic aftermath was severe. The Great Depression was deep in comparison with other European countries. In addition to that, after 1935 the National Socialist regime directed increasing portions of industrial capacity to military purposes[11].

The comparative delay in the development of high-volume consumer goods markets placed a severe constraint on the innovation opportunities for German firms, especially in comparison with their American counterparts. This was true particular for the high-volume machinery industries in which American firms achieved and sustained technological leadership throughout the period of the 1860s to World War II. Ludwig Loewe's voyages across the Atlantic made in the late 1860s in order to study the "American method" of manufacturing indicate that companies in the United States had already achieved technological leadership in the production of light machinery[12]. That leadership had initially developed in small arms production, was consolidated in the sewing-machine industry, and continued in the office machine, automobile, and electrical, as well as

9 Cf. note 7 for the source of data.

10 Cf. G. Bry, Wages in Germany, 1871–1945, Princeton 1960, pp. 266–322 for a comparison of Germany, Britain, and the United States. A French-German comparison is still lacking.

11 Cf. A. S. Milward, The German Economy at War, London 1965, p. 16.

12 For the interesting case of Ludwig Loewe see C. Matschoß, Geschichte der Ludwig Loewe & Co. Aktiengesellschaft, Berlin, 1869–1929, Berlin 1939, pp. 3–27.

in closely related machine-tool industries[13]. That leadership, it appears, was basically scale and market driven.

Edison's incandescent-lighting system, which Werner Siemens along with Emil Rathenau could admire for the first time at the Paris International Electrical Exhibition of 1881, provides good example to begin with. In developing incandescent lighting, Edison had all along (by autumn 1878 at latest) envisioned a system which would offer "light to all houses" – substituting gas lamps by supplying lighting at lower price[14]. In contrast to the type of arc lamps in use at that time the new incandescent lamp made possible low-intensity, small-space lighting[15]. This vision – to develop lighting for private household use and by doing this to open up a market of potentially gigantic scale – was obviously lacking at Siemens. There, experimental work had continued to concentrate on the improvement of the arc lamp[16]. This was a high-intensity lamp suited for large-space lighting of public places and buildings. This development was not directed to private households, but to administrative bodies and industrial customers. Siemens' omission eased the way for the entry into the high-voltage business of Emil Rathenau's German Edison Company (AEG), backed up by the Edison patents[17].

A similar technology gap later occured in the telephone industry. In that industry Siemens had to counter the entry of smaller firms, which in the second half of the 1880s began to manufacture multiple switch-boards under licence of Western Electric or imitated the Scribner construction. Subsequent improvement of manual switch-boards continued to draw on American technology. The technology of automatic switching which would rapidly be introduced in Germany during the interwar period had also been developed in the United States[18].

It appears that the underlying factors which created that gap were very much the same as in the example of the high-voltage power system. Amer-

---

[13] Cf. D. A Hounshell, From the American System to Mass Production. 1800–1932, Baltimore, London 1984, discusses the evolution of technolgy in these industries. Leadership is indicated by the massive transfer of technology to Germany, see n. 33 below.

[14] Cf. Th. P. Hughes, Networks of Power. Electrification in Western Society, 1880–1930, Baltimore 1983, pp. 28–33, quote p. 32.

[15] Cf. ibid., p. 31.

[16] The development of the arc lamp was advanced by the Head of the Construction Office, Friedrich Hefner-Alteneck, see F. Heintzenberg, Friedrich von Hefner-Alteneck, München 1951.

[17] Cf. J. Kocka, Siemens und der aufhaltsame Aufstieg der AEG, in: Tradition 17, 1972, pp. 125–42 on the relationship between Siemens and Rathenau's AEG.

[18] Cf. K.-H. Loesche/D. Leuthold, DeTeWe-Chronik, Berlin 1970, p. 36, p. 58 f. on Kellog-Switches, p. 134 on Siemens-Strowger construction.

ican inventors and entrepreneurs pioneered in recognizing the economic potential of an extensive telephone network which would reach to every single business and household[19]. Even though the German Postmaster General – responsible for telephony after it became a state monopoly in 1881 – had already expressed similar ideas in 1877 (inspired by an issue of "Scientific American"), administrative bodies would not make vigorous attempts to put them into effect during the decades to come. Throughout the 1880s and 1890s, the networks small size, its slow growth and the administrative policy of self-financing led to annual rates of charge which were so high that access to the system continued to be limited to the administration itself, business firms, and the upper strata of society[20].

A technology gap of the same sort later emerged in the electronics industry. Annual production of vacuum tubes indicates different industrial size and growth in the two countries. Whereas in the United States roughly 4.5 Million tubes were produced in 1923, in Germany it were only 0.1 Million. By 1931 annual output had risen to 52 Million in the United States, but had not exeeded 6 Million pieces in Germany by the same year[21]. By far the majority of the tubes were demanded by the rapidly growing consumer goods sectors.

A mass market for radio sets was first created in the United States. The concept of broadcasting was decisive for making radio a consumer product. This development was hampered in Germany by essentially the same factors as the use of the telephone. Broadcasting was also a state monopoly. As early as 1906 Emil Rathenau of AEG – which still had close ties to General Electric – had urged Hans Bredow – then Director of the AEG-Siemens joint venture for the development of radio-technology named "Telefunken" – to work towards a state of affairs in which "radio is needed in every home and through that mass production is made possible"[22]. But the state officials responsible thought that the American system of private broadcasting be "chaotic" and even some radio manufacturers were initially sceptical about it[23].

---

[19] Cf. R. Garnet, The Telephone Enterprise: The Evolution of the Bell System's Horizontal Structure, Baltimore 1985, pp. 77–89.

[20] Cf. F. Thomas, The Politics of Growth: The German Telephone System, in: R. Mayntz/Th. P. Hughes (eds.), The Development of Large Technical Systems, Frankfurt 1988, pp. 180–186.

[21] Cf. H. Petzold, Zur Entstehung der elektronischen Technologie in Deutschland und den USA, in: Geschichte und Gesellschaft 13, 1987, p. 347.

[22] Ibid., p. 348, n. 21.

[23] Cf. W. B. Lerg, Die Entstehung des Rundfunks in Deutschland, Frankfurt 1970, pp. 122–33.

Consequently mass production of radio sets was delayed, and at no point reached the scale to be found in America. By 1930 an estimated total of 12 Million households had their own radio set in the United States against roughly 3.2 Million in Germany[24]. As many as half of the 8 Million radio sets produced worldwide in 1931 were made in the United States[25]. As soon as their volume of production surpassed German output, American enterprises assumed technological leadership. As German companies scaled up production they drew increasingly on technology developed in the United States at such firms as AT&T and General Electric[26].

The automobile industry provides a variation on this theme. Here German pioneers such as Daimler had at first attained technological leadership in important areas. But again, continuous mass production started in the United States[27]. While Daimler and other automobile companies continued to focus on small-volume, but high-priced sectors, some German manufacturers of light machinery such as Adler, Opel or Wanderer did attempt production of small, low-priced automobiles. Throughout the interwar years, however, the volume of production continued to be very small compared to the United States. In 1929 managers of General Motors came to the conclusion, as Alfred P. Sloan reports, that the German domestic automobile market had reached about the same state of development as the United States market had attained in 1911[28]. Actually, conditions were even more backward. In 1903 the German automobile companies manufactured a total of 1.310 vehicles, compared to the total of 11,235 produced in the United States. By 1913 German output had grown to 17,162, far behind the U. S. figure of 461,500[29]. By 1928 United States production had increased to a gigantic 4.36 Million automobiles, but only 90,000 were manufactured in Germany – not even half as many as in Great Britain (212,000) or France (210,000)[30].

---

[24] For U. S. figures see: R. E. Summers/H. B. Summers, Broadcasting and the Public, Belmont/Calif. 1966, p. 45; for German figures see E. K. Fischer, Dokumente zur Geschichte des deutschen Rundfunks und Fernsehens, Göttingen 1957, p. 295.

[25] Cf. Petzold, p. 347.

[26] Cf. ibid., p. 364.

[27] Cf. D. A. Hounshell, From the American System, pp. 217–61; A. Kugler, Von der Werkstatt zum Fließband. Etappen der frühen Automobilproduktion in Deutschland, in: Geschichte und Gesellschaft 13, 1987, pp. 310–16 on Daimler.

[28] Cf. A. P. Sloan, My Years With General Motors, Garden City, New York 1964, p. 325.

[29] Cf. B. P. Bellon, Mercedes in Peace and War. German Automobile Workers, 1903–1945, New York 1990, p. 14, Tab. 1.

[30] Cf. Chandler, Scale and Scope, p. 209.

In the late 1920s Opel was the first German automobile company to start assembly-line production, technologically supported by General Motors which acquired an 80% interest in the firm in 1929[31]. The earlier and more dynamic growth of their domestic market enabled American enterprises to assume leadership, especially in the development of production technology and organizational know-how. This at least was what Fritz Opel and his engineers were looking for on their tours of the Ford and GM plants in the 1920s[32].

In this broad and significant range of high-volume machinery industries, the technological and organizational lead of American companies was the foundation for a massive transfer of technology (patents, licenses, machine-tools, know-how, man-power) and management techniques from the United States to Germany. This transfer began with Ludwig Loewe's introduction of the "American System" in the late 1860s and early 1870s, progressed through the waves of mechanization and rationalization which reached peaks around the turn of the century and in the late 1920s, and continued in the "Americanization" of West German industry after World War II[33].

The constraining effects of retarded market growth were not however, confined to the machinery and electrical industries. For the chemical firms too, the same sort of disadvantage became increasingly recognizable after World War I. This was especially true in the evolving synthetic materials industry. While Du Pont, for instance, made cellophane (a material invented in France) a significant product by developing a water-proof type for the needs of the large domestic food industry, it suffered from atrophy at IG Farben[34]. Again, in comparison with Du Pont, the development of synthetic fibers during the 1930s was hampered at IG Farben by market size and more directly on the organizational level by the lack of compe-

---

31 Cf. Sloan, p. 197, Kugler, p. 334, p. 136 ff.

32 Kugler, p. 334.

33 Cf. for instance H. Homburg, Anfänge des Taylorsystems in Deutschland vor dem Ersten Weltkrieg, in: Geschichte und Gesellschaft 4, 1978, on Taylorism; F. Blaich, Amerikanische Firmen in Deutschland, 1890–1918, Wiesbaden 1984, on imports of machine-tools; the contributions to Technikgeschichte 56, 1989, on rationalization; Th. P. Hughes, American Genesis, New York 1989, pp. 249–94 on Fordism and Taylorism; V. Berghahn, The Americanisation of West German Industry, 1945–1973, Oxford 1986, on the post-World War II period.

34 Cf. D. A. Hounshell/J. K. Smith, Science and Corporate Strategy. Du Pont R&D, 1902–1980, Cambridge/ Mass. 1988, pp. 170–80. On IG Farben see the chapter on cellophane in H. Voelcker, 75 Jahre Kalle, Wiesbaden-Biebrich 1938.

tence, experience and a suitable corporate structure to perform focused product development closely coordinated with market research and product advertising in the consumer goods sector[35]. German enterprises did undoubtedly have access to technological resources, but what limited their opportunities for innovation was the absence of a domestic consumer goods market as large and dynamically growing as the American one, and the lack of a corresponding vision, capability and structure to perform focused product development and aggressive marketing in such an environment.

2. *The Pattern of Success*

Weak innovative performance in the high-volume consumer goods industries contrasted with an obvious ability to innovate in low-volume, high-value-added market segments and niches of the capital goods sector. That ability was one major element on which the machinery and chemical firms built their international competitiveness. But it was not its only source. The basis of that competitiveness was of a threefold comparative advantage.

The first element of this was a cost advantage. Throughout the period of the late 19th to the middle of the 20th century wages continued to be low in Germany in comparison with the United States, Great Britain, and probably also France[36]. That advantage propelled the move into relatively labor-intensive, high-value-added, but low-volume sectors and niches. As a steel executive trenchantly put it in the 1920s: "We need to add value. With our still low wages we are the obvious country to add value"[37]. In other words, what continued to be the decisive comparative disadvantage for innovation and growth in the consumer goods sector, formed a significant advantage for the capital goods suppliers. That cost advantage could be converted into international competitive advantage to its fullest extent in the labor-intensive, low-volume, high-value added segments.

The second major advantage was an ample supply of skilled labor in the decisive phase of rapid corporate growth and organization-building of

---

[35] Cf. G. Plumpe, Die IG Farbenindustrie AG. Wirtschaft, Technik und Politik 1904–1945, Berlin 1990, p. 323, and part 3 below on organizational aspects.

[36] Cf. n. 10 above.

[37] Quoted in: G. D. Feldman/H. Homburg (eds.), Industrie und Inflation. Studien und Dokumente zur Politik der deutschen Unternehmer 1916–1923, Hamburg 1977, p. 226.

the late 19th and early 20th century. This was true for both blue- and white-collar occupations. The supply of skilled blue-collar workers had two principal institutional sources. The first was persisting handicraft traditions and establishments from which the machinery manufacturers continued to recruit a significant share of skilled labor throughout the 19th century. The second source was the so-called "dual system" of occupational training which took shape in the second half of the 19th century. Within this system the traditional apprenticeship was complemented by obligatory attendance at a vocational school. The supply of skilled labor, which also had a demographic dimension, was clearly an advantage in comparison with the United States. It is not yet possible to determine whether this was also an advantage in relation to other European countries. What is certain is that the attendance of vocational schools in connection with occupational training was made obligatory by law only in Germany[38].

The supply of scientifically and technically trained professionals and white-collar employees came from two major institutional sources. One was a dense, multi-layered system for technical education and training. That system consisted of technical secondary schools for the training of middle-ranked technical occupations (engineering-draughtsmen, calculators, etc.) and of an upper layer of technical institutes and universities. The system's origins reached back to the 1820s. In the second half of the 19th century it became increasingly more differentiated and sophisticated[39]. The second major institutional source was the traditional universities which had originated in the late 18th century and continued to perform the lion's share of training in the natural sciences. The whole process of institution-building was heavily supported, if not propelled, by the state, and to a far reaching extent financed through the public purse. It is still difficult to determine which institutional elements constituted a comparative advantages with respect to which country. It seems clear, however, that through the creation of technical institutes and universities in Ger-

38 On occupational training see G. Adelmann, Die berufliche Ausbildung und Weiterbildung in der deutschen Wirtschaft, 1871–1918, in: Zeitschrift für Unternehmensgeschichte, Beiheft 15, 1979, pp. 9–52; R. Schöfer, Berufsausbildung und Gewerbepolitik. Geschichte der Ausbildung in Deutschland, Frankfurt, New York 1981, esp. pp. 71–182 and K. W. Stratmann, Betriebliche Berufsausbildung, in: C. Berg (ed.), Handbuch der deutschen Bildungsgeschichte, Vol. IV, 1870–1918, München 1991, pp. 371–80.

39 Cf. J. Kocka, Bildung und soziale Ungleichheit: Entstehung und Differenzierung angestellter Mittelschichten im 19. und frühen 20. Jahrhundert, in: J. Kocka (ed.), Die Angestellten in der deutschen Geschichte, 1850–1980, Göttingen 1981, pp. 90–115.

many, the training of engineers caught up with or even surpassed French achievements, and became more advanced than in Great Britain[40]. In addition to this, the universities had reached relatively high standards during the 19th century and apparently continued to perform scientific training, despite the increasing numbers of graduates which made it more difficult to maintain the close integration of training and research.

The third and final major advantage, was the existence of a dense, mostly state financed research infrastructure within the traditional universities, newly created technical universities and numerous state supported research institutions, such as the institutes of the Kaiser Wilhelm Gesellschaft.

This particular configuration of comparative advantages and disadvantages which occured in the last third of the 19th century set the background for the strategic choice and organizational evolution of business enterprises. The strategic outcome was (1) a continuing move into high-value-added, low-volume segments and niches, in part as a reaction to the competitive superiority of western European and American firms in the high-volume industries, (2) a preference for product differentiation through technical quality, customization, service and short delivery periods and correspondingly little sympathy for price-leadership strategies, and (3) rapid diversification mainly through internal product development[41].

The large chemical and machinery firm – along with those of the metal industry – diversified early and strongly in comparison both to other German industries and their international competitors. In 1887 there were already 11 machinery and equipment manufacturers among the largest 100 enterprises which had more than 4 product lines, and even 4 firms which had more than 10 lines. Only in the metal industry had diversification made similar progress. Most of the largest chemical firms still covered less than 5 product lines, but it is worth noting that the dyestuff group in particular contained a very large number of different products – at Bayer there were about 200 different dyestuffs by 1886[42]. Diversification in

---

[40] Cf. G. Ahlström, Engineers and Industrial Growth, London, Canberra 1982; R. R. Locke, The End of the Practical Man, London 1984, pp. 29–88.

[41] German firms did, of course, engage in high-volume production. Most of the large enterprises performed both low-volume manufacturing of diverse, but technologically related finished products and high-volume production of vertically related commodities, semifinished goods and standardized components. But international competitiveness was clearly stronger in the low-volume, high-value-added sectors.

[42] Cf. J. Kocka/H. Siegrist, Die hundert größten deutschen Industrieunternehmen im späten 19. und frühen 20. Jahrhundert, in: N. Horn/J. Kocka (eds.), Recht und Entwicklung der Großunternehmen im 19. und frühen 20. Jahrhundert, Göttingen 1979, pp. 80–81, Tab. 1, and pp. 75–76 on Bayer.

other industries lagged far behind. Fragmentary evidence suggests that these enterprises were highly diversified, more so than those in other countries. French companies remained much smaller in size and far less diversified[43]. The majority of British firms, continued to be more specialized well into the interwar period[44]. So too did the largest American enterprises. Jürgen Kocka and Hannes Siegrist have shown that by 1907/09 and even by 1927/29 the largest German metal, machinery and chemical firms were clearly more strongly diversified than their American counterparts in the same industries[45]. Only in the electrical industry was the degree of diversification similar. While diversification slowed down in Germany during the interwar period, however, American enterprises diversified more agressively.

Diversification was achieved almost exclusiverly through internal product development. This accords with the pattern of German corporate growth more generally, merge with and acquisition playing only a minor role before World War I[46]. In the period between 1887 and 1907 only 23 out of a total of 150 mergers and acquisitions undertaken by 95 of the 100 largest companies of 1907 lead to significant diversification across the boundarics drawn by the census classifications of industries. More than half of them (12) were completed in the metallurgical sector[47]. The remaining ones took place in mining (4), chemicals (4) and electrical engineering (3). No diversifying merger or acquisition was to be found in the machinery industry[48]. Only during the interwar period, under conditions of economic instability and increased international competition, did diversification through merger and acquisition become more frequent, but less so in the machinery, electrical and chemical industry than in metals[49]. In

---

43 Cf. Levy-Leboyer, pp.. 119, 152–155.

44 This is also the impression from the data provided by A. D. Chandler, Scale and Scope, Appendices B.1-B.3.

45 For 1907/09 see J. Kocka, The Rise of the Modern Industrial Enterprise in Germany, in: Chandler/Daems (eds.), Managerial Hierarchies, p. 100 f., Tab. 3.6. For 1927/29 see H. Siegrist, Deutsche Großunternehmen vom späten 19. Jahrhundert bis zur Weimarer Republik. Integration, Diversifikation und Organisation bei den hundert größten deutschen Industrieunternehmen (1887–1927) in international vergleichender Perspektive, in: Geschichte und Gesellschaft 6, 1980, p. 82.

46 Cf. C. Huerkamp, Fusionen der 100 größten Unternehmen von 1907 zwischen 1887 und 1907, in: Horn/Kocka (eds.), Recht und Entwicklung, pp. 113–17.

47 Cf. ibid., p. 116.

48 Cf. ibid., p. 114, Tab. 6.

49 Cf. H. Siegrist, Strategie delle grandi imprese tedesche dopo la Prima guerra mondiale, in: P. Hertner/G. Mori (eds.), La transizione dall'economia di guerra all'economia di pace in Italia e in Germania dopo la Prima guerra mondiale, Bologna 1981, p. 299–300.

both Great Britain and the United States merger and acquisition were more important throughout[50].

This peculiar pattern of growth led to the evolution of a distinct type of corporate organization. Five organizational properties are particular noteworthy in our context. First, the enterprises reached a large size by the standards of that time. By 1907, the major chemical firms had about 7,500 to 9,000 employees. Siemens employed a total of 43,000 people, their competitor AEG roughly 31,000 in Germany alone. The equipment manufacturer Borsig had about 10,000 people on the payroll, and the highly diversified giant of the metal industry, Krupp, not less than 65,000[51].

Second, most of the large and rapidly growing enterprises soon adopted the managerial form of organization: ownership and control were separated and corporate government placed in the hands of salaried excecutives. This was similar to United States developments, but differed from the British path, where owners continued to play a central role in corporate management in many cases until after World War II[52].

Third, with the exception of Siemens, even the largest enterprises held on to a centralized corporate structure. Siemens was the only major German corporation to adopt a type of multidivisional structure before World War II. Here the creation of a corporate office followed the company's merger with the already large Schuckert enterprise in 1903[53] – a rather uncommon event in Germany before World War I. The predominance of the functionally centralized structure distinquished the German pattern from the American case where both horizontal combination and diversification through merger and acquisition eased the way for adopting the M-Form of organization.

---

[50] Cf. L. Hannah, The Rise of the Corporate Economy, 2nd ed., London, New York 1983, pp. 21–22, 74–76, 92–99 on Great Britain. For the United States in general see: R. L. Nelson, Merger Movements in American Industry, 1895–1956, Princeton 1959; N. R. Lamoreaux, The Great Merger Movement in American Business, 1895–1904, Cambridge/Mass. 1986, on horizontal combination; Chandler, Scale and Scope, p. 165 f. on diversification through merger. The acquisition of technology also played an important role in the diversification of Du Pont, see A. D. Chandler, Strategy and Structure, Cambridge/Mass. 1962, pp. 78–91 and the chapters in Hounshell/Smith, Science and Corporate Strategy. Again, it seems, the similarity was strongest between Germany and France, see: Levy-Leboyer, p. 124 ff.

[51] Cf. Kocka/Siegrist, pp. 106–112, Tab. 3.

[52] Cf. Chandler, Scale and Scope on all three countries, L. Hannah, Visible and Invisible Hands in Great Britain, in: Chandler/Daems (eds.), Managerial Hierarchies, pp. 53–56 on Britain and Kocka, The Rise of the Modern Industrial Enterprise, pp. 92–94, on Germany.

[53] Cf. Chandler, Scale and Scope, p. 467–471.

Fourth, growth through internal product development not only resulted in a high degree of spacial concentration of operational units, but also in the proliferation of strong technological interdependencies within firms. These interdependencies, as Alfred D. Chandler has argued, allowed the firms to exploit economies of scope[54]. On the other hand interdependencies intensified horizontal cooperation within organization, and hence lessened the possibility of control and coordination through the managerial hierarchy. Consequently, staff-offices (such as construction offices) or committees became more important for coordinating operations. In contrast, therefore, to the large managerial enterprises operating in scale-intensive industries which became so prominent in the United States, these organizations were distinguished by larger staff-units and shorter lines of command. In addition, technological interdependencies made it more difficult to establish detailed cost control and other forms of differentiated formal control systems. The stronger the technological interdependencies and the smaller the scale of operations, the more difficult and costly the handling of such systems became. Even though information systems based on cost accounting could be and indeed were established in the context of high-volume production of commodities, intermediates and standardized components, formal control of specialized operations and hence control of the technologically most dynamic organizational units remained relatively weak[55].

And finally, a range of factors tended to increase the power of the technical organizations and technologically trained executives within corporate structures and hierarchies, and to reduce the influence of non-technical professionals. These factors included the central significance of technological competence for implementing the firms' favoured corporate and competitive strategies: the technology-intensity of all major functional activities at this type of capital goods supplier including marketing; the minor significance of advertising and market research; the stemming of price competition through cartels, syndicates and other forms of inter-firm cooperation, and lastly the limited penetration of commercial control systems.

---

[54] Cf. ibid., p. 394, pp. 456–486.

[55] A fuller discussion of that subject can be found in: B. Dornseifer, Zur Bürokratisierung deutscher Unternehmen im späten 19. und frühen 20. Jahrhundert, in: Jahrbuch für Wirtschaftsgeschichte 1993, pp. 69–91. The difficulties of cost accounting in the face of operational interdependencies within the vertically integrated and diversified firm are reflected in the massive body of literature on accounting, organization, statistics and the like which proliferated in the late 19th and early 20th century.

The particular path of corporate development in Germany shaped by the configuration of comparative advantage and disadvantage in the late 19th and early 20th century did not lead to the coming of the type of managerial capitalism which emerged in the scale-intensive sectors of the American economy at that time. Nor did that evolutionary path allow a continuing dominance of commercial expertise and the persistence of a type of proprietary capitalism as it was the case in Great Britain[56]. The distinctive mark of the German pattern was the rise to dominance of technical expertise within corporations and the evolution of a type of technical managerial capitalism.

3. Technological Capabilities and Patterns of Innovation

What, then, were the organizational centres and functional sources of innovation at the most competitive firms? How was innovation operationally carried out? On what organizational capabilities did innovation rest and when, why and how did these capabilities evolve? These are the questions to be explored on the following pages by looking at product divisions of three major companies - Zeiss (scientific instruments), Siemens (electrical machinery and equipment) and Bayer/IG Farben (dyestuffs, synthetic materials).

3.1. The Product User as Inventor: Zeiss.

The strategic conduct and corporate evolution of Zeiss followed closely the lines identified as typical for the competitive cluster discussed above. Zeiss, however, began to diversify relatively late. Up to 1890 the firm, which had been set up by Carl Zeiss in 1846, continued to specialize in the manufacture of microscopes. Otto Schott's improvement of optical glass enabled it to penetrate the microscope market effectively during the 1880s. Expansion within that sector was followed by a phase of aggressive diversification: photographic lenses and gauges (1890), binoculars (1893), telescopes (1897), geodesic instruments (1908), and spectacles (1912)[57].

---

[56] The term 'proprietary capitalism' has been used by W. Lazonick, Organizational Integration, p. 80.

[57] Cf. H. Schultrich, Die Herausbildung des Industriephysikers im kapitalistischen Deutschland, dargestellt am Beispiel des Siemens- und des Zeiss-Konzerns, Ph.D. Dresden 1982, pp. 140–60.

This was achieved by diversifying the use of optical glass. This phase was succeeded by one in which the firm focused on rationalization and productivity increase. This had already started in the years before World War I, but was pursued more energetically during and after the war. By 1914 Zeiss had 3908 employees, a number which increased only slightly during the interwar years[58]. In 1912/13 the firm achieved 49 percent of turnover through exports. After the war the share of exports continued to be high (44 percent in 1924) and by 1929 had even surpassed the prewar level (54 percent). Between 1912 and 1929 turnover almost doubled[59].

At Zeiss' Microscope Division the matrix of innovation consisted of customers, in-house microscopy expertise which enabled the company to develop, diversify and test products, additional internal and external R&D capacity, and workshops. Customers continued to be a very important source of innovative ideas. The development of the stereomicroscope, for instance, was undertaken after suggestions made by the American zoologist Greenough; the non-chromatic condenser for microphotography rooted in ideas from Robert Koch; an apparatus for microscopic drawings was suggested by Professor Greil of the University of Innsbruck, and Professor Lehmann of the Technical University of Karlsruhe drew attention to the need for high-temperature microscopes[60]. Even though a quantitative measurement of the relative significance of customers as a source of innovation is not possible, these examples support Eric von Hippel's findings that the majority of innovations in the scientific instruments industry were initiated by customers[61]. The very close contact with the customer, the provision of technical services, and the support of scientific and medical research was thus not just a marketing and differentiation strategy, but was also an important way to gain innovative ideas.

The major internal source of innovation was the division's microscopy experts. The capacity to develop, diversify and test the use of microscopes was significantly strengthened in 1899 when the company hired the physicist Siedentopf (ultra-microscopy), the botanist Ambronn (polarization-microscopy) and the zoologist Köhler (ultraviolet-microscopy). All of them made significant contributions to product improvement and diversification. Köhler, who later headed the microphotographic unit of the

---

[58] Cf. ibid., p. 55, n. 8.
[59] Cf. ibid., p. 68, n. 3.
[60] Cf. ibid., p. 165.
[61] Cf. E. v. Hippel, The Sources of Innovation, New York, Oxford 1988, p. 4, Tab. 1–1.

Microscope Division, concentrated on the development of photographic attachments for microscopes which expanded the product's range of application and thus led to further diversification. Ambronn headed not only the Microscope Division but was awarded a chair in scientific microscopy at the nearby University of Halle – a chair which had been established on behalf of the company's former R&D director Ernst Abbe. Siedentopf attended to the development of ultra-microscopes, and in close cooperation with a chemist from the Schott Glass Laboratory achieved a significant success in 1902. Five years later he succeeded Ambronn in heading the division[62].

Around the turn of the century the division's internal organization was still informal and fluid. The tight integration of product development and marketing was most important. The division head had full responsibility for both product development and marketing (including pricing)[63]. He thus performed the functions of a product manager, however, unlike the general manager of a multidivisional firm, was not in charge of manufacturing. This form of organization favoured certain strategies, but constrained the pursuit of others. Without control of production costs the division head would hardly have favoured price leadership strategies. In order to achieve divisional growth the division head could attempt to penetrate markets on the basis of product differentiation. That option demanded continous technological improvement and adaptation to user needs. Alternatively he could force the division's further diversification, which basically meant diversification of range of application of the microscope. The drive towards product innovation was thus built into the division's form of organization. The continuing articulation of user needs by customers, and the company's microscopy experts, were critical to the pursuit of both strategies: product differentiation and diversification. What remained to be done was technological problem-solving.

In the process of product development the divisional staff cooperated closely with the centralized Construction Office on the one hand and the responsible Calculation Office on the other. The Construction Office carried out the design of the product's mechanical components, directed by an engineer. From there drawings and manufacturing instructions went directly to the mechanical workshop. The semi-centralized Calculation Offices concentrated on the design and technical development of optical components. The company had created three such offices by World War I.

---

62 Cf. Schultrich, p. 163.

63 Cf. ibid.

Each served one or more divisions[64]. In addition the calculation offices carried out long-term research projects on optical systems which were monitored by the company's Executive Board (*Vorstand*)[65]. In that sense the company had a centralized research unit. The calculation offices were directed by physicists. The Calculation Office serving the Microscope Division entered the process of product development in a routine manner by elaborating, in cooperation with the divisional staff, the basic outline of the specific optical system. The office went on to determine theoretically and arithmetically the system's optimal design. On that basis, and in close communication with the product division, the Contruction Office then designed a prototype. The prototype subsequently was tested by the divisional microscopy experts. If the prototype proved to come up to expectations the design was completed and the drawings finally passed over to the workshop.

In 1905 the Microscope Department employed four academically trained scientists. At least three of them were experts in microscopy. The R&D staff of the Photography and Binocular Division was about the same size[66]. Overall R&D capacity expanded as the company grew. By 1890 Zeiss had a total of 6 academically trained employees – most of them were physicists. By 1898 it was 9, 24 by 1903, and 51 by 1914. These were 1.6% of all employees in 1890, 1.9% in 1903 and 1.3% in 1914[67]. About two third of them worked at the divisions, while most of the others were located at the semi-centralized calculation offices[68].

There can be no doubt that the company's internal R&D staff, together with the customers, drove product innovation at Zeiss. However, it is equally as obvious that product innovation in the phase of rapid diversification could not have been carried out without a high level of technical skill and flexibility on the shop-floor. The higher the rate of technological change and the higher the demands for precision were, the more important was the skill and hence the education and occupational training of the workforce.

Well into the 1880s and 1890s, the machinery manufacturers' demand for skilled labor was served to a large extent by workers who had been trained at traditional handicraft establishments, but in-house occupational

---

[64] Cf. ibid., p. 173–174.

[65] Cf. ibid., p. 175.

[66] Cf. ibid., p. 161.

[67] Cf. ibid., p. 55, n. 8.

[68] In 1905 the divisions had a total of 16 academically trained employees; ibid., p. 161.

training became increasingly more important. At the large enterprises training was more and more separated from ongoing production processes and moved into special training workshops (*Lehrwerkstatt*). By the World War I, most of the major machinery manufacturers had established this kind of training workshops – as had Zeiss[69]. The technological capabilities of Zeiss thus did not rest on scientific competence alone, but on a particular symbiosis of science, handicraft traditions and state supported educational efforts.

3.2. Managing Complexity: Large-Scale Projects and Supplier Networks at Siemens.

The supply of heavy machinery, equipment and ready-made complex installations according to customer specifications was a particular strength of German enterprises which developed during the late 19th and early 20th centuries. Unfortunately almost nothing is known about this important subject. This analysis, using the example of Siemens serves more to explore the problem than to contribute to its solution.

Siemens' high-voltage division began to take shape around 1870. During the 1880s it rapidly gained organizational independence from the company's traditional low-voltage business. By the early 1890s most operations had been shifted to the emerging Werk Charlottenburg. This factory had thus become the core of an operationally autonomous division – containing capacity for manufacturing, marketing and R&D. In 1903 the high-voltage business was amalgamated with the Schuckert company which had been taken over and reorganized into Siemens-Schuckert Werke (SSW). Following the acquisition of Schuckert, Siemens centralized control and created a corporate office. By the early 1920s the division had become part of a multidivisional organization. The basic outline of that structure would remain unchanged until after World War II[70].

In the course of its history, the division went through four phases of organizational and technological evolution which each had its own centre of innovation. In the first phase, lasting from the middle of the 1860s to early 1870s, innovative activities centered on Werner Siemens who had set up the firm in 1847. During the 1860s, he developed a clear sense of the economic potential of the technology and in 1866 made a significant

---

[69] Cf. Adelmann, Berufliche Ausbildung, p. 26.

[70] Cf. Chandler, Scale and Scope, pp. 467–71.

contribution to its theoretical foundations by proposing the dynamo-electrical principle. Around 1870 innovative activities began to shift to the evolving Technical Department, in particular to the head of the Construction Office, the engineer Friedrich Hefner-Alteneck. In the first half of the 1870s Hefner-Alteneck focused on the development of dynamos, later turning his attention to the development of arc lamps. The dominance of the Technical Department lasted until about the mid-1880s. Innovative activities then began to shift to the rapidly growing multifunctional *Verkehrsabteilung* – the Department for External Communication. However, support by the division's Technical Department remained essential. A fourth phase began when the division endeavoured to increase efficiency. The first significant attempts in this direction were made around the turn of the century, but more comprehensive waves of rationalization occured shortly before, during and after World War I, in combination with strengthened central control. In that phase innovative efforts shifted to the manufacturing departments, leading finally to a more balanced distribution of innovative activities among the division's functional units. The phase of rapid expansion and diversification which began around the middle of the 1880s is the most interesting one in our context. Within a strikingly short period of time the division's *Verkehrsabteilung* developed an impressive capability to supply complex installations and to cope with multilateral supplier networks.

The *Verkehrsabteilung* consisted of commercial and technical units. The technical units performed five major functions: (a) technical marketing, (b) technical service, (c) project planning (*Projektierung*), (d) development of systems and components and (e) supervision of final installation (*Montage*). The department consisted of geographically decentralized branch offices (Technical Bureau) and corresponding functional units at the headquarters, at first called Technical Bureaus as well. There was frequent interchange of personnel between the external offices (*Außendienst*) and the headquarters (*Innendienst*). Werner Siemens had already created the nucleus of that type of multifunctional organization in 1868 when the company set up a Technical Bureau at its headquarters in order to cope with the technical correspondence related to the planning and installation of the large Indo-European Telegraph Line[71]. The office at the headquarters continued to serve the company's needs until 1890. Then the firm began to establish external Technical Bureaus and larger branch offices in Ger-

[71] Cf. Kocka, Unternehmensverwaltung und Angestelltenschaft am Beispiel Siemens 1847–1914, Stuttgart 1969, p. 136.

many and abroad[72]. By 1903 Siemens maintained 22 domestic and 8 foreign offices[73]. The staff of the external offices and the corresponding units at the headquarters was divisionalized along product lines and customer groups. The larger divisions had their own capacity for technological development (*Entwicklungsstellen*). There was close communication among the planning teams, the development staff and the division's central Technical Department.

By World War I the *Verkehrsabteilung* had grown into a huge organization able to market, plan, develop, install and service large and complex technological systems and single components. The number of employees indicates the large scale of the operation. In 1912 the entire high-voltage division had a total of 38,913 employees. 29,320 of these were counted as blue-collar workers, 7,185 as white-collar employees and 2,408 as other wage earners[74]. Of the 7,185 white-collar employees as many as 68% (4,883) worked in the *Verkehrsabteilung* alone – 42% (3,016) at the external technical and branch offices and 26% (1,867) at the Berlin headquarters[75]. A significant number of blue-collar workers must be added, although the exact number is uncertain.

At the headquarters the divisional planning staff cooperated closely with the division's (central) Technical Department. That department consisted of two major subdepartments: (a) the Construction Office (technical design and calculation) which also monitored the final testing of outgoing product (*Revision/Prüffeld*) and (b) the division's central R&D unit. The nucleus of a central research unit had been created in January 1873 when the physicist Oskar Frölich was made head of the company's laboratory. Up to then research had been the domain of Werner Siemens. As the company grew in size Werner Siemens' own experimental work gradually became more sporadic[76]. By 1875 Frölich had three laboratory technicians at his disposal[77]. In the course of the 1870s Frölich's lab was increasingly seen as a subunit of the evolving Technical Department directed by

---

[72] Cf. ibid., p. 369.

[73] Cf. Chandler, Scale and Scope, p. 465.

[74] The workforce of Siemens' low-voltage division (S&H) in contrast was not even half that size (17,967) and the proportion of white-collar employees was much smaller (2,909); see Kocka, Unternehmensverwaltung, p. 468. These differences indicate the high labor- and skill-intensity of the electrical machinery and equipment industry.

[75] Recalculated on the basis of the numbers in ibid., p. 469, n. 20.

[76] Cf. ibid., pp. 139–40. The first physicist had been employed at the company from 1856 to 1864, see Schultrich, p. 25.

[77] Cf. Kocka, ibid., p. 142.

Hefner-Alteneck. Frölich's work concentrated on the physics of low- and high-voltage electrical systems and machines and on the development of electrical measuring instruments[78]. By clarifying the physical foundations of electrical engineering Frölich contributed significantly to the firms capability to construct technological systems and their components in a purposeful, predictable and planned manner. He thus enhanced technological flexibility. In the course of reorganization following the acquisition of Schuckert in 1903, Siemens began to create a distinct corporate research department[79].

It was the planning teams, however, which were the centres of innovation in the phase of rapid expansion and diversification. These consisted overwhelmingly of engineers. Planning was directed by the division's most capable and experienced engineers and physicists, often recruited from the Technical Department. The planning process was essential for coordinating technical operations and for channeling the flow of information. Planning was performed in the context of networks which the project planners established between themselves, the customers, the Technical Department at the headquarters (especially the Construction Office) and cooperating and supplying firms. The management of these networks was an extremely complex and demanding task. As a result of the particular technological interests of the customers, the manufacturing departments (in standardization for instance) and the participating or supplying companies, the planning process took the shape of a multilateral negotiation process. Along with the increased "diversity of tasks", the history of AEG notes, "the demands put on every single person reached the limits of what it is possible to carry." Among the "responsible personalities" a "nervous strain" began to emerge[80]. In no other industry did technological innovation depended so heavily on managerial capability.

However, as in other parts of the machinery industry, innovation would not have been possible without the capacity of the workshop to cope with rapid technological change. The coordination of operations performed at the workshops was achieved through the Construction Office. The Construction Office designed the various components according to the specifications developed in the planning process. Single drawings then

[78] Cf. Schultrich, pp. 48–50.

[79] Cf. P. Erker, Die Verwissenschaftlichung der Industrie. Zur Geschichte der Industrieforschung in den europäischen und amerikanischen Elektrokonzernen 1890–1930, in: Zeitschrift für Unternehmensgeschichte 35, 1990, pp. 75–76.

[80] Allgemeine Elektrizitäts-Gesellschaft (ed.), 50 Jahre AEG, Berlin 1956. p. 83.

went to the proper (group) offices of the workshops. There tasks were allocated to the responsible master-artisans, delegated by them to the workers[81]. The work groups consisted of master-artisans, workers, and – with decreasing frequency – apprentices and were centred on certain work processes: forging, forming, turning, fitting, etc. Skilled workers were required because precision had to be achieved largely without making use of specialized machine-tools. Flexibility was necessary because the shape of components altered frequently. Capacity for flexible production, and hence the training and skill of both the blue- and white-collar work force, was a central dimension of Siemens' capability to innovate and compete in the heavy machinery and equipment industry.

3.3. Innovation as Routine: Dyestuffs and Synthetic Material at Bayer/IG Farben.

In the dyestuff industry innovation was carried out by the increasingly sophisticated and highly differentiated internal R&D organizations of individual firms. These organizations had gradually taken shape in the period from the 1860s – the decade in which most of the firms had been set up – to the 1890s. By the turn of the century innovation had become a highly routinized and rather profane activity – "scientific mass work" (*wissenschaftliche Massenarbeit*) as contemporaries called it[82]. In the decades to follow the basic outline and mode of operation of these organizations remained largely unchanged. Even in the course of reorganization following the merger of IG Farben no fundamental alterations occurred. To the contrary, responsible managers at IG Farben consciously did not disrupt the institutions which had grown during the preceding decades[83].

Product innovation was carried on by four interconnected sets of operational routines. Each set had a distinct strategic dimension. (1) Chemical analysis and synthesis, which were the decisive means to enter established markets (of "natural products") and hence significant for increasing diversification. (2) Systematic recombination of chemical compounds aimed at the increase of diversification within the boundaries of a given technologi-

[81] Cf. Kocka, ibid., pp. 344–46. A very instructive analysis of the shopfloor organization at Daimler-Benz has been provided by B. P. Bellon, Mercedes in Peace and War, pp. 25–60.

[82] Cf. G. Meyer-Thurow, The Industrialization of Invention: A Case Study from the German Chemical Industry, in: ISIS 73, 1982, p. 378.

[83] Cf. Chandler, Scale and Scope, p. 572.

cal field. (3) Selective testing and diversification of the range of application (*Anwendungstechnologie*). This so-called "application work" not only enhanced diversification but was also critical in achieving product differentiation. (4) "Scaling up" from the laboratory level to high-volume production (chemical engineering, *Verfahrenstechnik*), the operation central to the process of cost reduction.

In the dyestuffs industry, as in the case of scientific instruments, ability to apply results was at the heart of innovation. While it is widly accepted that chemical research was a *sine qua non* of innovation, the importance of its application or, more generally speaking, the function of the product user, should not be underrated. The major difference between Zeiss and Bayer was that the latter, along with the other dyestuff companies, not only vertically integrated the innovation function of the product user completely, but that they developed an ability to apply their results which was superior to that of their customers. That competence was also crucial for market penetration.

As competition increased the testing units grew into the largest and most powerful organizations within the R&D infrastructure. In 1896 only 36 out of 2,378 dyestuff compounds produced by Bayer's Main Laboratory passed first screening at the dyehouse (*Färberei*) and were subsequently tested on a larger scale. By 1906, 2,656 compounds were produced by the lab, 60 tested on a larger scale and 36 finally marketed[84]. From early on about half of the R&D personnel at Bayer concentrated on applications[85]. Besides testing, the colorists performed important marketing functions. By 1926 the application had become the single largest element of the R&D organization in the dyestuff industry. By then as many as 1,469 employees (including 291 chemists) were doing this type of work while 1,341 (including 384 chemists) were engaged in R&D at the laboratories[86]. Bayer's colourists thus performed a very similar function to that of the microscopy experts at Zeiss. The main difference was the degree of professionalization vis-a-vis the customers and the scale of operations.

The approach to R&D that Bayer had developed in the dyestuffs industry was later transferred to other fields such as pharmaceuticals (in the 1890s) and to artificial rubber (in the 1910s)[87]. However, after World War I,

84 Cf. Meyer-Thurow, p. 378, Tab. 7.

85 Cf. ibid., p. 365.

86 Cf. Plumpe, IG Farbenindustrie AG, p. 475, Tab.

87 For a description of R&D in the pharmaceuticals industry see: C. Duisberg, Von der Entwicklung und den Arbeitsmethoden der deutschen chemisch-pharmazeutischen Großin-

as high-volume consumer goods markets became more significant, this form of R&D organization and the way it was embedded into the overall corporate structure proved to be increasingly insufficient. This becomes obvious when we compare IG Farben with its major American rival, Du Pont.

In contrast to Du Pont, IG Farben did not create a centralized corporate research department. IG Farben's managers felt that such a department would create too wide a distance between research and other technical functions. Instead they favoured a close integration of R&D and production within the organizational framework of the "works" – frequently called the "technical organization". However, the lack of a centralized research department does not imply that IG Farben was not engaged in the type of research Du Pont's chemists performed at the corporation's central laboratory Polymer research undertaken by Wallace Carothers at Du Pont's Central Research Department is a striking case in point. At IG Farben the same strand of research evolved from research on solvents[88]. The scientific link between solvents and polymer research was acetylene chemistry. The new strand of research was then institutionalized. Until 1930 polymer research had been coordinated by the Solvent Committee. This committee then created a specialized Plastics Committee. The divisional Plastics Committee was later divided further along functional lines into the Plastics Scientific Committee, the Plastics Application Committee, and the Plastics and Rubber Production Committee[89].

The underlying reason why IG Farben did not create a centralized research department was the fact that the corporation, unlike Du Pont in 1921, did not adopt a multidivisional structure. Within the framework of the multidivisional structure a centalized research department was needed because the divisions focused on the development of their very product line and concentrated on the short-term increase of return on investment. The overall structure of IG Farben, in contrast, remained functionally centralized. Even though IG Farben's scale-intensive *Sparte* I (high-pressure and electro-chemistry) and *Sparte* III (Fibers, cellulose products) did come close to the type of autonomous product division, the critical elements of the multidivisional structure – a clear separation of operational

---

dustrie, in: C. Duisberg, Abhandlungen, Vorträge und Reden aus den Jahren 1922–1933, Berlin 1933, pp. 198–204.

88 Cf. Plumpe, p. 154 f.

89 Cf. ibid. and F. ter Meer, Technical Organization of IG Farben, Affidavit 23.IV.1947, NI-5186, National Archives, Record Group 238, T 301, Records of the Office of the United States Chief Counsel for War Crimes, Nürnberg, p. 4.

and corporate management, and the position of the general manager responsible for the development, production and marketing of one product line – were not introduced. The technical organization of *Sparte* II thus differed most strongly from the type of autonomous product division created at Du Pont. Containing the technological core of the predecessors (dyestuffs, pharmaceuticals, intermediates, etc.) it had the widest technological scope, the strongest technological interdependencies, the weakest control of costs and the largest organizational distance to its corresponding sales units (*Verkaufsgemeinschaften*). The technological scope of *Sparte* II was so wide that it alone covered half of the 10 product divisions Du Pont had created by the late 1930s[90].

In comparison with Du Pont, therefore, the modes of control differed most strongly at *Sparte* II. That was true with respect to both production and R&D. Elmer Bolton, Research Manager of Du Pont, got a clear impression of these differences when he visited IG Farben in June 1936:

"Dr. Bayer of the Central Laboratory of the Leverkusen plant [*Sparte* II] does not employ any project system, but is budgeted a certain amount of money each year which he allocates to the different lines of activity under his direction. On the other hand, Dr. Grimm of the High Pressure Laboratory at Oppau [*Sparte* I] is employing a project system somewhat similar to our own. Dr. Bayer stated that from time to time a review of the research results of his laboratory has been made in order to determine if the results justify the costs of the research, but in every case the management has been satisfied with the progress of the laboratory"[91].

These informal modes of control equipped the technical organization of *Sparte* II and its various units with a high degree of autonomy. DuPont's means of control were strongly formalized and the purpose of R&D was clearly defined. "The primary object of mechanical experimental work is directly or indirectly, to earn dividends for the stockholders of the Company. All work undertaken must, from inception to completition, be analysed and justified on the above basis." In order to get means "up to 500 Dollars" for instance, members of the Engineering Department had to turn in eight copies of the completed form No. 16822-Mechanical Experi-

---

[90] A more detailed comparison can be found in my unpubl. paper: Research, Innovation, and Corporate Strategy. Du Pont and IG Farben in Comparative Perspective, Harvard School of Business Administration, Spring 1989.

[91] Elmer K. Bolton to Jasper Crane, June 11th, 1936, Records of the E. I. du Pont de Nemours & Co., Hagley Museum and Library, Wilmington, Delaware, Series II, Part 2, Acc. £ 1038, p. 8.

ment Problem[92]. In contrast Carl Bosch of IG Farben made clear that "IG is not here to give big profits to its stockholders. Our pride and our duty is to work for those who come after us to establish the processes on which they will work"[93].

In comparison with Du Pont, IG Farben had an extremely "flat" managerial hierarchy even though the organization was much larger in size. The hierarchy of the technical organization did not have more than three or four layers: (1) Research group, (2) Laboratory Director (Department Head) and (3) Work Director who in the case of the main works was member of the Executive Board. By that time the hierarchy of Du Pont's R&D organization had more than double as many layers: (1) Research Groups, (2) Group Leader, (3) Research Supervisor, (4) Assistant Lab Director, (5) Lab Director, (6) Assistant Director (functional department), (7) Director (functional department), (8) General Manager (Product division), (9) Executive Committee, respectively in the case of the centralized Research Department: (10) Research Director who was also member of the Executive Committee[94]. Again, Elmer K. Bolton of Du Pont pointed out the major difference:

"There are about 100 chemists engaged in this laboratory [Oppau, high-pressure chemistry]. A rather unusual type of laboratory organization is employed. These chemists are divided into twenty-four divisions which report directly to Dr. Grimm [Lab Director]. The organisation does not have any Assistant Directors, but the Director has four Assistants, one of whom acts as an Assistant Director only in the absence of Dr. Grimm. [...] In all the laboratories of the IG which were visited it was very apparent that the Director took a very active part in the guidance of research work and delegated the administrative duties to others"[95].

The laboratories at Du Pont in contrast were much smaller in size, hence more specialized, and the R&D organization more strongly divisionalized[96].

The contrasting organizational structures of Du Pont and IG Farben led to different patterns of innovation. The organizational structure of

---

92 Cf. Standard Practice Book, Engineering Department 1920/21, Div. X, Du Pont Archive, Series II, Part 2, Acc. £ 1274, File 19A, p. 1A-1.

93 Quoted in P. Hayes, Industry and Ideology. IG Farben in the Nazi Era, Cambridge 1987, p. 3.

94 Cf. Hounshell/Smith, Science and Corporate Strategy, p. 306.

95 Bolton to Crane, ibid., p. 15, second part of quotation p. 8.

96 I think this was a general difference between German and American enterprises in the late 19th and early 20th century.

IG Farben allowed it to undertake exploratory research at the divisional laboratories. This mode of organization fostered crossfertilization, diversification and the exploitation of economies of scope in R&D. However, in comparison with Du Pont, it proved to be rather insufficient for focused product development. At Du Pont new and promising technology was quickly put into the hands of one single manager who was responsible for, and concentrated on, the development, production and marketing of his own product. Together with his functional staff he thus formed the nucleus of a new product division.

That was not the case at IG Farben. Even though the establishment of committees resembled the formation of product divisions in many respects, that mode of organization had two major flaws. First, responsibility was divided between various committee members or even different committees. Secondly, commercial marketing expertise was not represented in these committees. These were committees of IG Farben's technical organization, their members were technically trained and the organizational distance to the sales department was bridged only in a very weak manner. In addition, there was apparently much tension at this interface between the technical organization and the centralized sales units at the Frankfurt headquarters. The extraordinary coherence of IG Farben's technical functions was paid for by an organizational segmentation of technical operations on the one hand and the commercial activities, most importantly sales, on the other. This structure cultivated technological virtuosity, but was insufficient for focused product management, especially in high-volume consumer goods markets.

4. *Summary*

In the industrial development of Germany strength and weakness in the area of innovation were placed side by side. From the late 19th to the middle of the 20th century German enterprises displayed a noticable weakness to innovate in the modern, high-volume consumer goods industries when compared with American firms. This weakness resulted from the comparative delay in the development of high-volume consumer goods markets. It was however contrasted by a striking strength to innovate in low-volume, high-value-added capital goods industries. That strength was an essential part of the international competitiveness of firms, but it was not its only source. International competitiveness also rested on a threefold comparative advantage: relatively low costs of labor, ample supply of skilled labor, and access to highly developed, state supported scientific and technical institutions. That particular configuration of comparative advantages and disadvantages which occured in the late 19th and early 20th century set the pattern for the selection of competitive and corporate strategies and led to the evolution of a type of technical managerial capitalism in Germany.

The innovative activities of the internationally competitive firms were directed towards enhanced product differentiation and diversification. Within that strategic context the function of the product user and the application of results had a particular significance for innovation at all three companies which have been examined. In order to meet the technological requirements articulated by in-house experts and customers the firms developed different routines and technological capabilities. While Bayer/IG Farben endeavoured to meet these requirements by increasing the scale of R&D operations, Zeiss and Siemens developed the capacity for flexible production.

IV.
The Formation and Development of Industrial Research and Technology

Introduction

NATHAN ROSENBERG

I would like to call your attention at the outset to the considerable degree of present-day relevance of the subject of these papers. Obviously, relevance is not a necessary condition for the study of economic history. Nevertheless, where such relevance exists, it is at least noteworthy. For each of these papers demonstrates a concern with issues of great policy relevance today, not only inside the nascent European Community but in all countries that share a concern with the conditions of future industrial progress. These are: (1) How to exploit the growing potentialities of scientific knowledge in the cause of economic progress? (2) What is the optimal form for organizing the research process in different industries? Cooperation vs. competition. And, [closely connected with (2)] (3) What are the appropriate roles of public vs. private R&D?

The industrial research laboratory was, itself, one of the great innovations of the 19th century. But it is also an institution that has played distinctly different roles in different industries and in different national environments. One of the important contributions of these three papers is that they provide extensive documentation for this point in the interwar years.

Each of these papers deals with the emergence of research capabilities in the interwar period. The specific stories focus on a variety of initiating or driving forces. In Norway, as Sejersted shows, the driving force was not some pre-existing technological or scientific capability, but the opportunities offered by the availability of cheap hydroelectric power. In Germany, as Erker shows, the "long-standing tradition of state dominated research" provided a fertile base upon which the electrical industry would build a powerful industrial research base by 1930, when there were no less than 840 laboratories and institutes for industrial research. In the Netherlands, by contrast, until the early 1930s, "[...] all industrial research was carried out in the emerging research laboratories of the Dutch 'Great

Five': Philips, AZKO, Unilever, Shell and DSM" (Erker). With the use of patent data, Cantwell shows the persisting comparative advantage of continental Europe in the chemical sector in the interwar years – offering a sharp contrast with Great Britain which remained tied to the older technologies of textiles and shipbuilding.

These papers also confirm the view that, in contrast with the USA, industrial research laboratories developed later and more slowly in Europe. Nevertheless, the papers also underline the extent to which German performance surpassed those of the rest of Europe, especially in the critical chemical and electrical industries. But the papers also confirm the view that must have been prominent in the minds of the organizers of this conference, that the interwar period was one of extensive growth and development in those research institutions upon which improved economic performance was coming increasingly to depend. European industrial research capabilities were vastly greater at the beginning of the Second World War than they had been at the end of the First.

All three papers look carefully at the role played by structural change in the formation of science-based industries. In this sense, they show a greater degree of continuity with chapter 3 than might have been expected. This may represent a happy coincidence. A more plausible hypothesis is that the organizers of this conference were extremely shrewd and far-sighted in their organizing functions. They are to be congratulated. The interwar years has long been an underdeveloped and neglected period in scholarly research. These papers represent important first step in overcoming that neglect.

The Choice between Competition and Cooperation: Research and Development in the Electrical Industry in Germany and the Netherlands, 1920–1936

PAUL ERKER

1. Introduction

Industrial R&D played a major role in the development of technological rivalries between nations, as well as in the ideas concerning the creation and maintenance of wealth and prosperity which characterized the interwar period[1]. There was a remarkable increase in levels of industrial research, which became more and more 'internalized' and institutionalized. The establishment and expansion of central research units within companies had already become significant throughout the industrial world, but exactly what role did R&D play in firm strategy with respect to the changing nature of competition? Markets were unstable and rapidly changing. A major aim of R&D-intensive corporations was thus to control the dynamics of technical developments and to reduce economic uncertainty in an attempt to stabilize their continually endangered technological positions. Also significant is the way in which economic and technological change affected intrafirm research management and research work, changing the role, organizational structure and indeed the very nature of intrafirm laboratories.

This paper will discuss corporate R&D using a comparison of the German and Dutch electrical industries, specifically the firms of Siemens,

---

[1] Studies on the history of R&D in European Industry are conspicuous by their absence. For a recent example see P. Erker, Die Verwissenschaftlichung der Industrie. Zur Geschichte der Industrieforschung in den europäischen und amerikanischen Elektrokonzernen 1890–1930, in: Zeitschrift für Unternehmensgeschichte 35, 1990, p. 73 ff.

AEG, Telefunken, Osram and Philips. Each company had a different technological profile and environment for innovation. From this emerges the question of how these factors affected the nature of their reaction to competition, and whether they adopted an offensive or defensive orientation towards research strategy, R&D structure, and innovation policy. The first section will look briefly at the national pattern of industrial R&D in electrical engineering in Germany and the Netherlands, with a brief outline of British developments[2]. In Germany a complex network of R&D linking the major four enterprises had to be adapted to the rapidly changing technological environment. Shadows of the old rivalry between AEG and Siemens, together with technological convergence complicated the arrangement of industrial research. On the other hand, in the Netherlands, where the backing of a strong national research system was lacking, electrotechnical R&D was centred within the Philips company itself, which had already become a multinational firm. The main focus lay on the most rapidly changing technologies in electrical engineering, which were the then 'high-tech' products: electric light and radio. The main part of the paper will examine more closely the inter- and intrafirm relations within German R&D, especially at the 'Forschungslaboratorium' (FL), the central research unit, at Siemens. The 'Natuurkundig Laboratorium' (Nat. Lab.) at Philips will be compared in detail with its counterpart at Siemens. The leading questions for both intrafirm research systems will be how was the problem of centralized vs. decentralized research dealt with, and who took the decisions and provided the impulse for change? Finally what were the research traditions and dominant research concepts? The fluctuations in competition and cooperation between Philips and the German firms will be considered against the background of market domination and cartelization, with some concluding remarks about their technological and research positions on the eve of World War II.

2. *"Networks of Science": State, Cooperative and Private Research. The European Pattern of Industrial Research*

In the 1920s American observers of the scientific scene in Europe noticed the prominent role of the state in promoting and coordinating industrial

---

[2] A study analysing intrafirm R&D in one of the British electrical companies is still needed.

research[3]. The First World War had not only induced the "cult of science" but had also brought state participation in research to the fore. In the interwar period nearly all European countries tried to establish a close "network of science" linking government organizations, universities, cooperative research associations, and private industrial research institutions, conforming to guidelines established by central "national research plans". But the rate of expansion, the pattern of organization, and the operating capacities of these networks varied significantly from nation to nation. In Germany, with its long-standing tradition of state-dominated research, the strongest network of industrial research developed, mainly motivated by the aim of regaining the prewar world market position. The Physikalisch-Technische-Reichsanstalt (PTR) and the Kaiser-Wilhelm-Gesellschaft with its 40 research institutes, were the central research organizations in the industrial field. These were well functioning institutions, supported by both government and industry but which remained independent in order to conduct fundamental as well as applied research[4]. Large corporations in the chemical and electrical industries, as well as other corporations such as the Vereinigte Stahlwerke, Krupp, Zeiss and MAN, had established their own research and testing laboratories since the turn of the century. Where plants were small, or where cartels, communities of interest, or trade associations existed, Germany also developed a good many cooperative research associations. This was especially true in the textile industry, where the Deutsche Forschungsstätte für Textilstoffe und Textilindustrie, joint organizations founded in 1918 and 1920 in Krefeld, Karlsruhe, and Dresden, were well known and successful examples of cooperative research. The institutionalisation of German industrial research laboratories in the interwar period thus emerged both on an intrafirm level and within collaborative research organisations. The total number of scientific and technical research organisations grew from 232 in 1919, to 840 laboratories and institutes for industrial research in 1930[5]. Close linkages between

---

[3] See M. Holland, Research in Europe. A Comparative Study of the National and Industrial Organization, Paper presented before a meeting of the Division of Engineering and Industrial Research of National Research Council, New York, November 21, 1924.

[4] See D. L. Cahan, An Institute for an Empire. The Physikalisch-Technische Reichsanstalt, 1871–1918, Cambridge/Mass. 1989, and L. Burchardt, Wissenschaftspolitik im Wilhelminischen Deutschland. Vorgeschichte, Gründung und Aufbau der Kaiser-Wilhelm-Gesellschaft zur Förderung der Wissenschaften, Göttingen 1975.

[5] Cf. VDI (ed.), Deutsche Forschungsstätten technischer Arbeit. Handbuch der auf dem Gebiet der Technik und verwandten Wissenszweige arbeitenden Forschungs-, Versuchs- und Prüfanstalten und dergleichen, sowie der diese Anstalten unterstützenden Vereine, Körperschaften und Organisationen, Berlin 1919, and Die Technisch-Wissenschaftlichen Forschungsanstalten, Berlin 1931.

these three research traditions guaranteed coordination in research and scientific exchange in German industry. This mixture of government sponsored industrial cooperative research, fundamental research in universities, and industrial research in largescale industries, backed by a highly developed educational system, constituted the German strength in industrial research until the mid-1930s.

In Great Britain, Germany's major European rival in the application of science to industry, both government research activities and the intrafirm implementation of research laboratories were far less developed. The National Physical Laboratory was a research organization similar to the German PTR, but its links with industry were weak. It was only in 1916, with the founding of the Department of Scientific and Industrial Research (DSIR) that a central coordinating agency for industrial research was conceived. With a commitment to foster cooperative as well as private research activities in industry and to coordinate this with a "national research plan" the DSIR had to contend with both a relatively "non-research-minded" British industry and with its own policy of "less Government in Business"[6]. The main stimulatus to cooperative research came with the inauguration of 22 industrial research associations, in such fields as the automobile, textile and electrical industries. But from the start the long-term financing of these associations was insecure, and problems arose from the different research needs and expectations of small and large firms. The Electrical Research Association, for example, was a cumbersome, bureaucratic organization with 30 to 50 subcommittees, each with 150 people, responsible for seeking and defining technical problems, assessing their importance, guiding scientists, and studying the application of the results of research. The new spirit of research was too weak to bring about any fundamental change to the 'research mindedness' of small firms where techniques had persisted for generations. The largescale firms such as GEC, MetroVickers, and British-Thompson-Houston (BTH) were likely to be more interested in conducting their own research than getting involved in collaborative projects. But here too, research organization and its intrafirm implementation were far behind German competitors. In addition, most of the British electrical industry had been established with a great deal of financial assistance from foreign investors, under whose in-

---

[6] Cf. Holland, Research in Europe, p. 2, M. Sanderson, Research and the firm in British Industry, 1919–1939, in: Science Studies 2, 1972, pp. 107–51 and for the following I. Varcoe, Cooperative Research Associations in British Industry, 1918–34, in: Minerva 19, 1981, pp. 433–63.

fluence it remained. Therefore, British firms suffered from technological and economic domination from abroad. The industry's R&D in the interwar period continued to be carried out largely in the United States[7]. Finally, each of the components of the British network of research, government research institutions, research associations, and private company R&D laboratories, were weak, and lacked the strong backing of scientific education at the university level. The whole coordination of science showed remarkable defects which produced a lag in industrial research, placing the UK behind Germany[8].

The Netherlands, as an industrial late-comer and taking into account both its small size and its scarce scientifically-trained personal, showed the greatest deviation from the European pattern of industrial research. Until the early 1930s, all industrial research was carried out in the emerging research laboratories of the Dutch "Great Five": Philips, AZKO, Unilever, Shell, and DSM. Nevertheless, the absence of a large home market and of the necessary human resources fostered the early move of Dutch companies towards multinational activities and an extensive transfer of know-how from abroad. For Dutch industry, research had, from the beginning, gone beyond national boundaries; cooperative research, therefore, meant the coordination of the R&D activities with subsidiary-companies abroad as well as with competing firms in Europe and the USA. Only in 1930/31 did the first outlines of a national research network emerge. The university system intensified its training of scientists and engineers. The greatest change came as the result of a law passed in October of 1930 establishing a strong central organization for applied research called the National Council for Scientific Research (TNO), (later supplemented by a council for pure scientific research) with a subcommmittee for the coordination and stimulation of industrial research. Expansion also occurred

---

[7] Cf. A. D. Chandler, Scale and Scope. The Dynamics of Industrial Capitalism, Cambridge/Mass. 1990, pp. 354 ff. Moreover the British electrical industry underwent, in the late 1920s, a process of concentration through mergers and acquisitions which brought serious difficulties to the creation of new "research synergies", e. g. in 1929 between the laboratories of BTH and MetroVickers to form one research unit for the new Allied Electrical Industry (AEI).

[8] Current research in Great Britain tends to weaken this argument. In comparison to France or Italy, indeed, the British industrial research system was far ahead. Cf. A. Russo, Science and Industry in Italy between the two world wars in: Historical Studies in the Physical Sciences 16, 1986, pp. 281–320 and T. Shinn, The Genesis of French Industrial Research 1880–1940 in: Social Science Information 19, 1980, pp. 607–40 as well as R. Thorstendal, Industrial Research and Researchers in Sweden 1880–1940 in: ibid., pp. 641–61.

in the number of industrial R&D-laboratories in the Netherlands in the interwar period which rose from about 30 in the early 1920s to 95 in 1939. This did not change the dominant role of intrafirm research organisations in the major Dutch firms[9].

As in other industries, the technological push of the First World War and the expansion of industrial R&D changed the markets and the conditions of competition for the electrical industry. New competitors emerged and a stream of new products promoted industrial diversification, accelerating the rate of innovation. For the electrical industry, the only way to cope with this situation seemed to be an intensification of its R&D activities. But between the wars, there was a certain innovation-split in electrical engineering technologies. With the high-voltage technology (electric light and power), product and process innovations had temporarily come to an end. An intricate net of agreements, concerning the exchange of patents, licenses, experience, and technological competence connected all the main companies in the world[10]. Low-voltage technology (radio and consumer electronics), however, was marked by a huge potential for innovation. Both the electric light industry, based on gas-discharge and fluorescence technologies, and the radio and telecommunication industry, based on the electron-tube and valve technologies, were influenced by the slow but fundamental change in basic science from the older electron physics to quantum and solid state physics. This was the field which formed the mainstay of industrial research. In these fields an invention could suddenly create new markets for industry. In the same way, marketing areas which were still profitable and carrying heavy investments could quickly be flooded and destroyed. Thus the 1920s and 1930s represented a period of uncertainty and instability in the electrical engineering field, as far as the choice of investments, developments and marketing activities where concerned. But in research the field was much clearer. Thus research

---

[9] Cf. J. J. Hutter, Nederlandse Laboratoria 1860–1940, een kwantitatief Overzicht, in: Tijdschrift for de Geschiedenis, Geneskunde, Natuurwetenschaften, Wiskunde en Technik 9, 1986, pp. 150–74 and E. Bloemen, Onderzoek en ontwikkeling door het bedrijfsleven in Nederland, 1919–1939, Philips Company Archives (PCA) unpubl. paper 1977.

[10] Cf. B. Kiesewetter, Internationale Verflechtung der Elektroindustrie, in: Deutsche Volkswirtschaft 23, 1934, pp. 724–29, V. Schröter, Die deutsche Industrie auf dem Weltmarkt 1928–1933, Frankfurt 1984 and P. Czada, Die Berliner Elektroindustrie in der Weimarer Zeit, Berlin 1969. For the technological developements see A. Bright, The Electrical Lamp Industry. Technical Change and Economic Development from 1800 to 1947, New York 1949 and R. MacLaurin, Invention and Innovation in the Radio Industry, New York 1949.

managers and laboratory directors in industry were working to strengthen the new technological base which promised to be the route for the future development and expansion of the industry.

3. Strategy and Structure of Industrial Research: Siemens, AEG and the Problem of Technological Convergence

Siemens and AEG, though both profiting from science, had contrasting concepts about its place in industry and their research strategies differed fundamentally. Pursuing the whole spectrum of electrical engineering, Siemens to a great extent exploited the technical inventions of its own physicists and ingenious founders. AEG, however, engaged mostly in high-voltage-technology, concentrated on the exploitation of foreign patents. Until the late 1920s small development departments, closely related to the manufacturing processes were run in order to adjust the patents to their own production lines. Though one of the greatest and most innovative producers of incandescent and gas-discharge-lamps, AEG contributed little research of its own. Whereas Siemens, which was in close contact with the PTR and the Technical University of Berlin, conducted its own research, Erich Rathenau's AEG succeeded in developing outside inventions into technically sound products on a large scale[11]. But by the mid-1920s the shift in focus from the high-voltage field to the low-voltage sector as the leading edge of technological development forced AEG to revise its strategy. In a painful process the whole company underwent a remarkable change from a marketing- and development-orientation to a research-orientation, moving decisively towards low-voltage products. Whether further technological excellence could be achieved and maintained through reliance on foreign know-how or by intrafirm R&D had become a matter of debate at AEG in the early 1920s. Several attempts to establish a central research laboratory had found no support from the heads of the company, and only in 1928, with the coming of Hermann Bücher as head of the executive board, was the *Forschungsinstitut der AEG* established[12].

---

[11] Cf. Wissenschaft in Berlin, Berlin (O) 1979 and W. Fischer, The Role of Science and Technology in the Economic Development of Modern Germany, in: P. Lundgreen (ed.), Zum Verhältnis von Wissenschaft und Technik, Bielefeld 1976, pp. 210 ff.

[12] For AEG see C. Ramsauer, in: Jahrbuch des Forschungsinstituts der AEG 1, 1929, pp. 401 f. and ibid., Zehn Jahre AEG-Forschungsinstitut, in: AEG-Mitteilungen, April 1938, pp. 275 ff.

The AEG laboratory, headed by the experimental physicist Prof. Carl Ramsauer, rapidly expanded from 5 to 11 research departments[13], and by 1933 had doubled its number of scientists from 30 to about 60. Ramsauer pursued new concepts of both managing R&D and of its organizational links with the company as a whole. Work in the *Forschungsinstitut* was aimed primarily at basic research[14]. Free from the responsibility for the day-to-day manufacturing and developmental problems of the plants, the central laboratory of AEG, through "pure research" and "technical research", sought to secure a new technological basis for the company. This goal survived the Great Depression and was only slightly modified by the reduction in research investment and personnel which accompanied it. The main problem faced by the *Forschungsinstitut*, however, was how to survive as a central research unit in a heavily decentralised company. Each plant at AEG constituted a separate organisational unit which was self-sufficient in the conduct of all internal operations[15]. Organising the flow of scientific results from the central laboratory to these units posed difficulties. Ramsauer circumvented the problem, which, as we will see later, also confronted Siemens, in two ways. First, a number of research departments were transferred to the manufacturing units once they had finished the basic research for a product or process innovation. In this manner, during the course of the 1930s, the Isolation Laboratory, the Sound-Film Laboratory and the Low-Voltage Laboratory became parts of their corresponding works[16]. It was only a short term solution, however, because such transfers undermined the capabilities of the central research laboratory in the long-run. A second way to manage the problem was the establishment of the *Technisch-Physikalischen Werkstätten*, whose task was to test the output of the *Forschungsinstitut* for manufacturing on a large scale, and to improve product maturity in order to accelerate the transfer to the works. Finally, in 1933 Ramsauer became a member of the board of management, thus helping to give weight to the interests of the "research community" in the company's policies.

---

[13] Allg. Physik/Atomphysik, Technische Physik/Elektronenoptik, Akustik, Röhren, Elektrotechnik, Wärmetechnik, Allg. Chemie, Isolierstoff-Chemie, Metallurgie, Magnetismus und Spezielle Aufgaben, see Jahrbuch des AEG-Forschungsinstituts 1, 1929, pp. 402 ff.

[14] Cf. C. Ramsauer, Sollen die technisch-physikalischen Forschungslaboratorien der Industrie rein wissenschaftliche Forschung betreiben?, in: Zeitschrift für technische Physik 10, 1929, pp. 223–26.

[15] Cf. R. Brady, The Rationalization movement in German Industry, Berkeley 1933, p. 178, and Chandler, Scale and Scope, p. 546.

[16] Cf. C. Ramsauer, Aus dem Forschungsinstitut der AEG, in: AEG-Mitteilungen 1935, p. 28 ff.

AEG, though heavily hit by the economic depression in the mid-1930s, became a research-minded company with an effectively organised and, as far as the link between pure science and manufacturing was concerned, a well-balanced research and development unit. The *Forschungsinstitut* represented the type of modern research laboratory established at the end of the 1920s not only in the electrical industry but in others as well. In contrast to the older electrical laboratories, which had been established at the turn of the century, and which concentrated on electric-lamp technology, the *Forschungsinstitut* was concerned instead with the new field of electronics. The advantage enjoyed by AEG's management was that in order to change the company's technological basis, they did not have to overcome a rigid research organization possessing obsolete technological know-how. With a hint of envy, Prof. Hans Gerdien, the head of the central research laboratory at Siemens, noted at the end of 1930, that "the whole system of production of the AEG *Forschungsinstitut* is, in a certain sense, contrary to that of our *Forschunglaboratorium*. Tests, which in our opinion were closely connected to practical application, were only done to a relatively small extent. It surpassed work, that without essential change could also have been produced by a university institute or the Kaiser-Wilhelm-Gesellschaft research institutes"[17].

Indeed, the central research laboratory at Siemens not only represented a different research strategy but also a different research tradition. Since the late 1890s Siemens had pursued intrafirm research with an organizational philosophy of strict separation between central research, serving equally both Siemens and Halske (S&H, primarily telegraph and telephone) and Siemens-Schuckert (SSW, primarily electric light and power), and departmental development. In the second half of the 1920s, the departments for research and development had expanded as rapidly as the rest of the company. Research work had passed to a great extent to the *Zentrallaboratorium* (ZL) in the Werner-Works Z (WWZ) of S&H, to the *Meßgerätelaboratorium* (Technical Laboratory) of Werner Works M, and to the *Abteilungsspezìallabor* (Special Department Laboratory) at Werner Works F (WWF), which were counterparts of the central *Forschungslaborato-*

[17] "Die ganze Arbeitsrichtung des Forschungsinstituts der AEG [steht] in einem gewissen Gegensatz zu unserem Forschungslaboratorium. Untersuchungen, die in unserem Sinne enger an die praktische Verwendung angeschlossen sind, werden nur in verhältnismäßig kleinem Umfang ausgeführt. Es überwiegen Arbeiten, die ohne wesentliche Änderung auch aus einem Universitätsinstitut oder etwa einem Forschungsinstitut der Kaiser-Wilhelm-Gesellschaft hervorgegangen sein könnten". SAA 15/Le 405, Jahresbericht des FL 1929/30, pp. 56 f.

rium (FL)[18]. With growing corporate decentralization, divisionalized research had come to outweigh centralized research. But the task of the FL "[...] had mainly shifted to 'practical research' in close connection with the needs of the works and departments, [and] in comparison scientifically-oriented research work decreased"[19]. The FL had also expanded from 3 to 24 research departments, which in 1929 had more than 200 scientists and engineers[20]. The problem faced by the Siemens management in the 1920s not only concerned the inter- and intrafirm structure of the research system, but also the blurred contours between research and development, which also made a revision of the financing and the control of the research activities necessary.

Despite the early institutionalization of R&D at Siemens, the FL had always had to struggle for intrafirm acceptance. Gerdien, without being a member of the board of management, had difficulty promoting his organization to the directors of the works who were scarcely research-minded. To promote the transfer of research results, Gerdien and his scientists went into the manufacturing units, hoping "[...] to give the directors of the works and departments impulses for new research work and to develop a closer bond between the FL and the production works".[21]. But rapid technological development had led to overlap of research fields that had previously been strictly separated between the FL and ZL. This resulted in duplication of research work. In addition, the SSW works in particular heavily criticized the research work of the FL,

[18] The official names for the research laboratories at Siemens are misleading. The central research laboratory for the whole company only was called "Forschungslaboratorium", whereas the departmental laboratory of the Werner Works Z at S&H carried the name "Zentrallaboratorium".

[19] "[...] besteht wesentlich aus "praktischer Forschung" in enger Fühlung mit den Bedürfnissen der Werke und Abteilungen, demgegenüber Forschungsarbeiten einer mehr wissenschaftlich auf weitere Ziele abgestellten Richtung zurücktreten". Hans Gerdien, Ziele und Aufgaben technisch-physikalischer Forschungsinstitute in der Industrie, in: Zeitschrift für technische Physik 10, 1929, pp. 218–22, and Jahresbericht FL von 1929/30, p. 56, in: SAA 15/Le 405.

[20] The Physikalische Abtlg., Physikalisch-technische Abtlg., Wärmetechn., Kältetechn., Magnetische Abtlg., Elektronenröhren-Abtlg., Röntgen, Mechanischtechn. and Metallographische, 2 physikalisch-chemische Abtlg., one Chemisch-techn., Anorganisch-chemische, organisch-chemische Abtlg., the Kolloidchemische Abtlg. and the Konstruktionsbüro, the Mechanikerwerkstatt and the glastechnisches Labor. Cf. Jahresberichte des FL 1923/24, p. 1, in: SAA 11/Lb 63 (NL Haller).

[21] "[...] den Herren Werks- und Abteilungsleitern Anregungen für neue Fortschrittsarbeiten zu geben und das Zusammenarbeiten des Forschungslabors mit den Werken und Abteilungen enger zu gestalten". SAA 15/Le 405, Jahresbericht FL von 1929/30, p. 9.

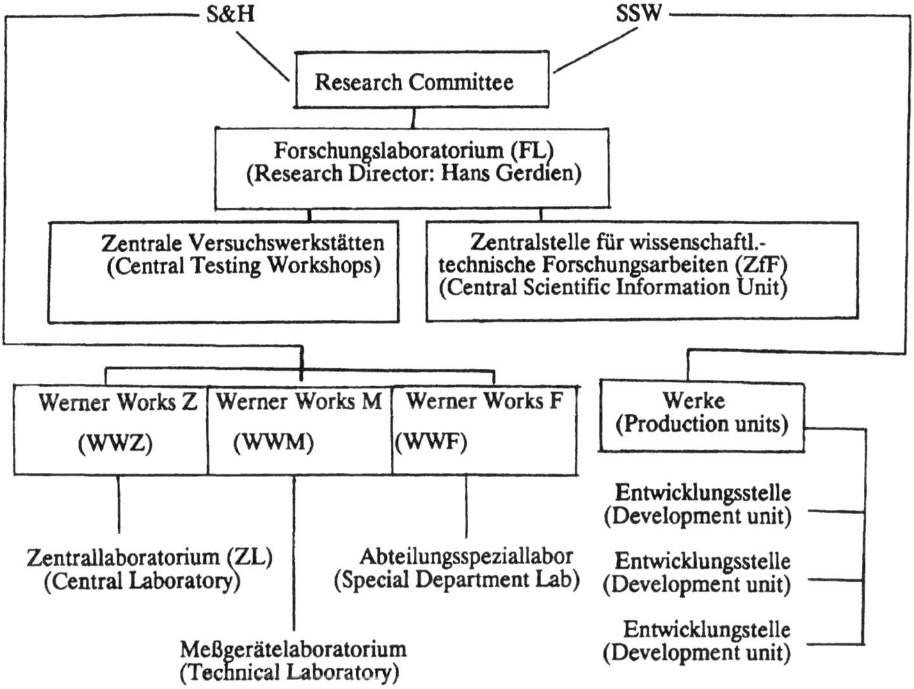

Figure 1: Organization of R&D at Siemens (1930)
Source: SAA 11/Li 321 (v. Buol); Siemens-Zeitschrift 17 (1937), p. 113.

both for its alleged expensive and its "pure" and unfeasible character [22]. Originally, the initiative for new research came both from the manufacturing units and the FL itself, coordinated by the ZfF (*Zentralstelle für wissenschaftlich-technische Forschungsarbeiten*). But in 1926, all research undertaken by the FL still had to be approved by the works and departments, in terms of both the research objectives and funding[23]. A committee of managers, consisting of S&H and SSW directors, had the final decision. In spite of these structures of the research units Siemens still invested heavily in industrial science. Its R&D budget rose from about 3 Mio.RM in 1922 to about 16 Mio. RM in 1936[24]. While sales doubled the R&D budget rose five times as much reaching about 2.5 percent of sales for the whole concern but 5 percent for S&H alone. About 80 percent of all R&D money went on development work.

---

[22] Cf. the conference of the board of management in 19.3.1927, in: SAA 11/Lf 262 (NL Köttgen).

[23] Cf. the reports of the ZfF in: SAA 15/Lg 963.

[24] Cf. SAA 11/Lf 262 (NL Köttgen), 15/Le 405 and 49/La 581.

The Great Depression nearly dealt a final blow to the poorly coordinated research system at Siemens. FL as well as ZL suffered sharp cuts both in R&D funding and research personnel. More serious, however, was the fact that the intrafirm transfer of know-how was blocked, so that the whole system of research financing broke down. "Existing fruitful relationships were terminated from all sides, Gerdien complained in his annual report 1930/31. The works and departments that up to this point had already been reluctant to approve resources for our work, reduced them to nothing within a short time"[25]. He desperately tried to emphasise the more promising results of research work to counteract the threatened transfer of the remaining research projects into the specialised laboratories of the production units. Marketing and product orientation threatened to supercede the research orientation of the Siemens company.

Faced with such major problems the board of management succeeded in carrying through a reorganisation and reform of the company's research system[26]. The testing workshops of the FL and ZL were amalgamated, the different research fields were coordinated, merged, and redistributed to the different research laboratories. The FL, for example, now had to conduct all physical research, the ZL of WWF did all semiconductor research, and the ZL of WWZ concentrated on the whole spectrum of theoretical and applied chemistry. An annual research budget replaced project financing, thus giving the scientists the leeway they needed for long-term research. The basic problem of technological convergence within the Siemens research system had been solved, but the blurring between research and development, between basic and pure research and technical and applied research, and between central and developmental laboratories still existed, and in 1935 another reorganisation took place. Central research was divided into FL I under Gerdien and FL II under the newly hired physicist Gustav Hertz. The FL II then became the intrafirm centre of basic science and research, concentrating on the new quantum, solid state, and atomic physics. Despite the internal problems with regard to the strategy and structure of industrial research, by the mid-1930s Siemens had established a closely coordinated but still relatively flexible R&D system.

---

25 "Es wurden uns von allen Seiten die bestehenden fruchtbaren Beziehungen gekündigt [...] Die Werke und Abteilungen, die bisher schon sehr zurückhaltend in der Bewilligung von Mitteln für unsere Arbeit gewesen waren, reduzieren diese nun innerhalb kurzer Zeit auf Null". SAA 15/Le 405, Jahresbericht FL 1930/31, p. 9.

26 Cf. SAA 11/Li 321 (NL von Buol).

The nature of competition between the two giants of the German electrical-equipment industry varied from product line to product line. In some – equipment for power, light, streetcars, and railways – the two competed functionally and strategically for market share. In other lines, such as incandescent lamps, cables, and some types of turbines, they cooperated through conventions or joint sales syndicates. In still others, such as wireless telegraphy and high-speed electric locomotives, the two formed joint ventures. The choice between competition and cooperation reflected more the nature of the product and the existing market than the predilection of either company for one or another type of interfirm relationship. The more standardised the product was, and the less susceptible to rapid technological innovation, the greater the likelihood of cartelisation[27]. The choice between competition and cooperation also affected R&D. In some areas the firms competed in research, while in others they formed strategic research alliances. Siemens and AEG even made plans to establish an extensive pooling of their whole research activities to reduce R&D costs. The practical problem with this research policy can best be seen in connection with Osram and Telefunken, both jointly founded by Siemens and AEG.

Before World War I Siemens and AEG had pooled their research and patents in communications technology through the newly established Telefunken GmbH (under this name from 1923). In 1920 both also sponsored the foundation of Osram GmbH to coordinate their immense scientific knowledge and experience in electric bulb physics, and to conduct their formerly competitive electric bulb business. Both joint ventures became powerful companies in the interwar period, representing very different types of technology-based and research-oriented organisations. Telefunken was conceived as an "invention company" (*Erfindergesellschaft*) with an elaborate research branch. The scientific basis was the huge reservoir of patents from the parent companies, upon which further R&D activities of Telefunken were run. In high-frequency, low-frequency, television, and electron-tube laboratories, together with three developmental laboratories, both fundamental and applied research work was carried out[28]. Because manufacturing had remained at Siemens and AEG and only a marketing branch supplemented the company organisation, there

[27] Cf. Chandler, Scale and Scope, p. 472 f. and W. Feldenkirchen, Zur Unternehmenspolitik des Hauses Siemens in der Zwischenkriegszeit, in: Zeitschrift für Unternehmensgeschichte 33, 1988, p. 25 ff.

[28] Cf. O. Suhr, Telefunken, in: Der deutsche Volkswirt, 15.11.1935, p. 319 f.

was no need to choose between centralised or decentralized research. Nevertheless the continuous flow of research results between the subsidiary and the parent companies posed problems. In view of the rapid innovation cycles in communications technology, a real coordination and pooling of research meant that more and more research fields had to be transferred from the research departments at Siemens and AEG to Telefunken, thus supporting their technological foundation but at the same time weakening their research capabilities. In 1925 Telefunken, fortified with important parts of the electro-optical research from Siemens and AEG, began to undertake television research, and in 1931 took over the electro-acoustic laboratories of the parent companies. Nevertheless technological change and technological convergence led more and more to overlapping and duplication of research work. Know-how transfer and research coordination failed, fostering research competition between Telefunken on the one hand and Siemens and AEG on the other hand. Siemens especially displayed deep discontent about the work of the Telefunken laboratories, while at the same time there were growing conflicts with AEG about its share of research inputs and manufacturing output.

Osram, in contrast to Telefunken, had mainly been conceived as a manufacturing and marketing company whose research capacity was reduced to developmental work and defensive "patent research." A small research laboratory, the Studiengesellschaft für elektrische Leuchtröhren mbH, had to a large degree focused on imitation and adaptation of foreign technical improvements in lamp and tube technology. The high degree of cartelization of the whole electric bulb industry also gave little incentive to intensify basic research. Between 1925 and 1934, Osram's research budget was cut from 1 Mio. RM to about 0,5 Mio. RM[29]. Conflicts also emerged in this area between Siemens and AEG over the interpretation of patent law[30]. Behind the debate in November 1932 over compensation for their respective contribution of patents and research the two big firms also struggled for technical as well as research and scientific domination. Siemens again criticised the inadequate research activities of Osram, complaining that it did not pay enough attention to technological progress, a charge that was justified in the case of research into gas discharge and fluorescent lamp technology, which Osram was too late and too hesitant in pursuing.

Thus, a complex network of scientific and industrial R&D developed between the big four German electrical companies, which functioned in

---

[29] Cf. SAA 15/Lm 367.

[30] Cf. SAA 11/Lg 171.

an often haphazard manner. In 1924, in the electron tube research field, in addition to the tube laboratory of Siemens, there was a competing development laboratory for electron valves at Osram, but under the technical leadership of the Telefunken valve laboratory. With the retirement of its head researcher in 1927, the Telefunken laboratory became affiliated to the Osram laboratory, but it was later separated again, only to be unified in 1934 for a second time, this time under the auspices of Telefunken. In contrast to the rapid technological progress in electrical engineering, which required vigorous research and a quick shift from research to development and testing work, the German electrical research network reacted slowly.

In 1929 the head researcher of the Siemens valve laboratory, Walter Schottky, painted a very gloomy picture when he made a critical inventory of the research situation. The R&D work was disjointed without any responsible coordinating unit. If there were research findings or ideas that could lead to the development of a new type of radio valve, the missing connection with the manufacturing works hampered the impulse to innovate. "In all fields, a closer cooperation between the valve researching, -developing and valve manufacturing units within the Siemens-, AEG-, Telefunken- and Osram-group is necessary in order to succeed against the foreign competitors", Schottky concluded[31]. In 1931 Siemens' and AEG's management openly stated that Telefunken, with its present technological and scientific capabilities was no longer a viable entity. Attempts by Siemens to transfer its radio laboratories to the Telefunken research unit failed because of the poor condition of the Telefunken laboratories and dissent from AEG. This situation, as well the monopoly in the home market and the high cartelization in the international radio industry, hampered technical progress. Reluctantly, and far later than its European and American competitors, Telefunken changed its technical thinking and research work from radio component to radio system development[32]. But

---

[31] "Auf allen Gebieten wird eine engere Zusammenarbeit als bisher zwischen den röhrenverwendenden und röhrenerzeugenden und entwickelnden Stellen innerhalb der Siemens-, AEG-, Telefunken- und Osram-Gruppe notwendig sein, um die Entwicklung in dem Tempo zu fördern, wie es gegenüber der Auslandskonkurrenz notwendig scheint". SAA 11/ Lg 722 vom 18.2.1929, p. 5.

[32] An exception was the television valve research at Telefunken. Despite the depression, great sums had been invested in the new electronic picture-transmission-technology. Expanding the research work in the television laboratory from a handful of scientists to over 300 researcher in 1938 with an annual budget of 1,4 Mio. RM, Telefunken – with great support from the new NS regime – for a while had a technological edge on its competitors.

the main factor behind the "Telefunken problem" was technical convergence in communications technology. The old separation of the working and research fields of Telefunken and its parent companies ceased; the former high-frequency and low-frequency research converged into one great carrier-frequency technology for communication engineering (*Nachrichtentechnik*). The highly divided industrial research organisation for what was a rapidly converging technology in the interwar period made Siemens realize only in the late 1930s that "[...] in its feeling for a research tradition and company heredity it seems to be necessary to take over again all the technology Telefunken was working on" [33].

At the end of the interwar period, therefore, the complex network of cooperation, characterising so much the industrial research in the German electrical industry, seemed to have come to an end. Above all, Siemens reasserted its own technological strength and research potential and ultimately was unwilling to yield this. The negotiations undertaken in 1930/31 with AEG for a global understanding, which would have included close cooperation in research and a technological fusion in order to reduce high R&D expenses, sank into oblivion. Indeed, during these negotiations, the principals envisaged a common committee for R&D to work out joint directions for industrial research at Siemens and AEG[34]. but they failed to produce an agreement. Because of the very different research traditions and R&D cultures of the two great electrical companies, they had virtually no chance of succeeding, even under the threat of competition from abroad.

4. The Natuurkundig Laboratorium (Nat.Lab.) at Philips: Research for a World Market

As with the other electrical companies, the initial years of industrial research at the Dutch company of Philips, founded in 1891, were dominated by the urgency to safeguard the company's market position as a producer of incandescent lamps. Until 1914 Philips carried out development work only in order to imitate and adapt foreign technology. The First World War, however, promoted intrafirm research at Philips more than for any

[33] "[...] im Gefühl seiner Verantwortung für ein überkommenes Erbe es unabweisbar [erscheint], daß es die gesamte zur Zeit von Telefunken bearbeitete Technik wieder in die eigene Hand übernimmt". SAA 4/Lf 720, Memo of 20.8.1940.

[34] Cf. SAA 4/Lf 530.

other company in electrical engineering. Dutch neutrality allowed Philips to concentrate its research activities on the new light technology and in the whole area of consumer electronics at an early point[35]. To offset the apparent disadvantage of being on the outside of the main stream of technological development carried out for war purposes, the Philips company was forced to become self-sufficient in research. Because it was not expending any effort on defence contracts it was able to develop and market some of the world's first radio tubes for public use, and its researchers spent time studying gas discharge phenomena which gave Philips a significant edge over its competitors when the war was over. In the early postwar years, Philips succeeded in dominating the Eastern European markets of Poland, Hungary, Czechoslovakia, Yugoslavia, and Romania, whereas in Germany and most of the other Western European countries, market entrance was hampered by the hurdle of patents and cartel quotas[36]. Thus, right from the start, Philips's motive for industrial research and its goal for the *Natuurkundig Laboratorium*, founded in 1914, was to overcome these hurdles and to pave the way to transnational expansion of manufacturing and marketing facilities. The firm's development into a multinational company owed much to its research. Although the real organization of R&D at Philips was only slightly different than that of others in the electrical industry, the company's research strategy and its corporate culture were fundamentally different from those of the German firms.

Despite the early institutionalisation of the *Natuurkundig Laboratorium* as a central research unit, the organizational and conceptual framework of the industrial research system at Philips was not fully established until 1924, when the laboratory moved into its own building[37]. At that time 100 scientists and engineers worked in five research departments on light/gas discharge, radio/acoustics, chemistry/metallurgy, x-ray and mathematics/fundamental physics, expanding in 1934 to 370 employees. Gilles Holst, the head of the central laboratory, perceived its task primarily as one of fundamental research. Since the laboratory's establishment, he had

---

[35] Cf. A. Heerding, Geschiedenis van de N. V. Philips Gloeilampenfabrieken, Part II: Een Onderneming van vele markten Thuis (1891–1922), Leiden 1986.

[36] Cf. F. Philips, 45 years with Philips, Dorset 1976, pp. 22 ff.

[37] Cf. for the following: M. Voorhove, Over de Organisatie van wetenschapplijk onderzoek, Philips Company Archives (PCA), mimeographed (1928), A. Garrat, The Story of the Philips Laboratory at Eindhoven, 1914–1946. The Reign of Gilles Holst, 2 volumes, PCA, mimeographed (1977) and "Veertig Jare van het Natuurkundig Laboratorium", PCA, mimeographed (1955).

insisted that it must be totally independent and should never become a trouble-shooting organization for the production departments. This strict separation of research from development, or as Holst expressed it, of "scientific research" from "technical research"[38]. had its repercussions in the financial and managerial sections of the company. Holst got his research funds directly from the board of management, and he was also responsible to it. In 1926 he himself became a member of the board.

At Philips, in contrast to Siemens or AEG, nearly all of the initiative and ideas for new research fields and projects came from the *Natuurkundig Laboratorium.* Thus, even in the early 1920s, along with its research work on radio valves and gas discharge tubes, the central laboratory pursued new product lines, whereas the board still backed the established light technology. "A condition for the success of research," Holst once stated, "has always been the wish of the company to market new products without being deterred by the fact that a new product did not fit into the existing line of manufacture. This philosophy has always strongly influenced the way of working in our laboratories. When a branch road promised better results than the main road we did not hesitate to take the branch road"[39]. Holst's comment about the research strategy at Philips is essentially true, but it was, in fact, the other way around: the *Natuurkundig Laboratorium* and the research community not only influenced but essentially determined the company's policy.

Applied or technical research at Philips was conducted in the so-called Chemical Lab.V and in some test departments under the same roof as the manufacturing divisions. Two other links between research, development, and production distinguished Philips during this period. Holst deliberately released older scientists from the central laboratory to become development men in the factories, transferring their experience and skill from research to manufacturing and applied technology[40]. Holst also employed the so-called *Proeffabriek* (testing or pilot factory), comparable to AEG's *Physicalisch-Technische Werkstätten* and Siemens' *Versuchswerkstätten.* It was a special department established to find out how new devices which had originated in the *Natuurkundig Laboratorium* could best be manufactured and to set up a small-scale production line to build them. In this way research could anticipate the problems the factories would have with new

---

[38] Cf. G. Holst, Industrielaboratorien, in: Physica. Neederlandsch Tijdschrift voor Natuurkunde 2, 1930.

[39] Quoted from Garrat, Vol. I, p. 150.

[40] Cf. ibid., p. 170.

products or production methods. The *Proeffabriek* not only served as a meeting place between research and production, it also allowed small-scale production of items in which the manufacturing divisions were not interested[41]. So the test factory spared the central laboratory from solving development problems, and essentially helped to maintain Holst's concept of separating R from D, thus avoiding the problems faced by Siemens.

In the course of the 1920s, Philips' research underwent a remarkable diversification, considerably changing its focus. Radio research was expanded significantly, concentrating on tube and rectifier research as well as developing a theoretical background in carrier-frequency physics. In 1925 the *Natuurkundig Laboratorium's* research in light bulbs had almost ceased, and researchers began to explore new fields. Thus, in 1927, the company initiated television research, and in 1930 it entered the telephone field. This decision was made in full awareness of the economic depression. The board of managers felt that the Great Depression might have a serious effect on the radio market, and they were looking for another market, less subject to economic fluctuations. Telephony was a commodity business which they felt should offer stable prospects, although from a commercial point of view it did not look immediately attractive, and from a scientific perspective it was a highly competitive market, especially for a technological latecomer like Philips[42].

Throughout the *Natuurkundig Laboratorium* influenced or significantly determined the company's innovation policy, backing the decisions of the board of management or pointing to the direction that Philips needed to pursue in the near future. The laboratory played an especially important role in the making of two major decisions, whose importance and full significance were certainly not realized at the time. First, in the early 1920s, the company shifted from component thinking to system thinking with regard to communications technology, especially in radio research, and with this shift moved from producing highly developed components to manufacturing complete systems[43]. This choice of technological systems brought a more differentiated view of the objects of research and their scientific interdependences, thus helping to manage the problem of technological convergence that their German competitors were confronted with. The second fundamental decision, forced by Holst's laboratory in 1930, and thus some years earlier than at Siemens, was to move Philips

[41] Cf. ibid., pp. 175 ff.

[42] Cf. Garrat, Vol.II, pp. 114 ff.

[43] Cf. Garrat, Vol.I, pp. 156 ff.

into the new field of solid state physics and semiconductor research. Thus the *Natuurkundig Laboratorium* had become more and more a force in establishing company policy at the highest level. At this time, all the drive for new products and innovative aims came from it; it was a clear case of research and technology push: industry pull[44].

In the shadow of the depression, however, Philips also undertook a reorganisation of its research system. The long-established patent department had developed into a coordinating and controlling unit for research matters, comparable to the ZfF at Siemens. Intrafirm transfer of know-how was rationalised, and a committee for evaluating research results was established[45]. The reorganisation only slightly affected the organizational structure of the *Natuurkundig Laboratorium*. However, in line with the reorganisation at Siemens in 1930, the mechanism for financing research at Philips was also rationalised. Never previously exposed to financial pressures from the manufacturing departments, the central laboratory had only intermittently received money for its annual research projects from the board of management. After 1930, the laboratories received a lump sum annually, without any strings attached. The amount of research money was fixed at about one percent of the company's turnover. Thus the independence of the *Natuurkundig Laboratorium* was institutionalised, and this move was strengthened by the correlation of the research budget to the firm's overall earnings[46]. The most fundamental change was in the development research departments. Affected by the depression, the work of the *Proeffabriek* declined. On one side its functions were taken over by different departments which began to strengthen or set up their own development laboratories, and on the other a Central Workshop was established in 1935, which took over many of the former services of the test factory.

In the mid-1930s, Philips had a highly developed system of industrial research, which showed a well-balanced and discrete organisation of centralised and departmental research similar to that at AEG. In this system, the *Natuurkundig Laboratorium* succeeded in defending its philosophy of never becoming subordinate to the factory and involved exessively in development work. With the guiding hand and driving force of the central laboratory, the Philips Company underwent a fundamental change from an imitating incandescent lamp manufacturer to a highly diversified and innovative electrical engineering firm.

---

[44] Cf. ibid., p. 161.

[45] Cf. ibdi., pp. 181 f.

[46] Cf. ibid., pp. 204 f.

5. The Key to Prosperity? Industrial Research and the Technological Drive on the Eve of World War II

After World War I, as far as electrical engineering was concerned, there was a broad development and use of existing technologies[47]. The rapidly expanding markets for light bulbs and radio valves gave the electrical industry a sound basis for prosperity. At the same time, in research laboratories in industry as well as in the universities, a remarkable scientific push preceded the Great Depression, leading to the invention and development of products and technologies that helped to fuel the postwar boom of the 1940s and 1950s. Semiconductor and solid state research as well as product and process developments in television and magnetic recording were established during the interwar period.

Each of these technologies, partly overlapping or converging in their emerging product lines, required a special organisation and management of the intrafirm R&D process. In the electric light sector, defensive improvement work prevailed, whereas for electronic tubes R&D led to dramatic changes. In solid state physics, however, still in its infancy, R&D centred on fundamental research, with scientists attempting to understand its theoretical foundations. Through cartels and marketing and price agreements between competitors, however, the markets for the advanced technologies of the interwar period developed a rather rigid structure. In the early 1920s, AEG and GE on the one hand, and Siemens and Westinghouse on the other had come to an agreement about sharing their patents and technological know how. During the following years these agreements were extended to more and more American as well as European electrical firms, regulating the influence of each on the market of the others. But patent pooling and tightening of the marketing regions did not insulate firms against market fluctuations or relieve them of the rising cost of R&D work. There was a peculiar swing between competition and cooperation in the electrical industry which provided the setting for industrial research. After 1921, for example, a cartel of "The Big Four" in the radio industry (Marconi, CSF, RCA, and Telefunken) existed which relied upon a close exchange of patents and research results. But it did not prevent Telefunken from being confronted with Philips's aggressive strategy of penetrating and dominating the German radio market[48]. In 1925, Telefunken succeeded in signing a contract with the Dutch company, which pre-

---

[47] After World War II the electronics industry faced nearly the same situation.

[48] Cf. SAA 4/Lf 720 vom 29.3.1928 for the negotiations between Telefunken and Philips.

vented Philips from entering the German radio tube market. However, in 1929 Philips found a loophole through which to enter the market. Intense competition did not, however, preclude Philips and Telefunken from coming to an understanding on how to eliminate other companies from the European tube market. They did this by developing and producing jointly a cheap tube, which succeeded in suppressing the activities of smaller or new tube companies[49]. In 1931 Philips and Telefunken signed a basic contract specifying close technological cooperation, but at the same time each firm had procured exclusive rights to new valve patents from RCA, which gave rise to further competition and conflicts.

In 1922 Osram, Philips and GE reached an agreement over the electric bulb market that bound them to exchange patents or technical knowledge and to respect each other's home markets. Two years later this agreement was institutionalised in an electric light cartel (called Phoebus)[50]. Nevertheless, rivalries emerged between Philips and Osram over business leadership. Thus in 1933/34 a deep conflict broke out about mutual rights and duties with regard to patents on the new gas discharge lamps[51]. Philips, having developed this innovation and scaled its production up for a mass market, put pressure on the entire European lamp market, arguing that the new technology did not fall under the cartel regulations. Despite prolonged negotiations the firms in this cartel never reached a new agreement on market sharing. Applied research to reduce manufacturing costs, to circumvent patent barriers, and to avoid the pressure to harmonize their own products to those of the competitors had an important function in this cartel poker. Marketing calculations directed the flow and progress of technology, and not competition in the research field.

On the eve of World War II, looking back to 1914, Philips, AEG, and Siemens had taken the same technological path to a large extent, but each had employed different means to master it. For each of them, the nature of expanded, diversified, and decentralised corporations made research a far more sophisticated undertaking than it had been earlier. Industry had become more closely tied to the universities, recruiting had become more problematic, managing researchers had become more complex, the routine of research had become more formalised, and coordinating research had become more challenging. The main questions they all confronted centered on linking: How should companies link technology with strategy,

---

[49] Cf. Berliner Börsencourier of 1.2.1930.

[50] Cf. Bright, p. 132 and Schröter, pp. 260 ff.

[51] Cf. in detail SAA 11/Lg 171.

and how should managers best link the R&D laboratories with the rest of their organisation? As we have seen, Philips and the German electrical companies came up with different answers which seem to have had little to do with national patterns but more with the respective technological traditions and foundations of the firms themselves. At Siemens, which was engaged right from the start in the whole field of electrical engineering, a deeply rooted and broad-based research tradition determined the strategy. At AEG, at first heavily engaged in high-voltage technology and then entering low-voltage technology, the actual technological centre of gravity determined the company's strategy. At Osram and Telefunken, company policy determined the research strategy, which was based on a single technology, thereby fostering the tendency for both companies to survive on their manufacturing and marketing functions and to rely upon others for new knowledge and new technology. At Philips, which broadened its technological foundation from one technology to the whole branch of consumer electronics, it was just the opposite. Here, the research strategy determined the company's overall strategy and policy. Siemens and AEG tried to concentrate on the new technological and scientific fields by separating certain branches of development and applied research; Philips succeeded in maintaining its concept of strict intrafirm separation of research on the one hand, and development on the other, thus having at its disposal a more flexible system of industrial research. These differences aside, the R&D systems of all the firms discussed here overcame the years of economic depression relatively intact, although they had been partly reorganized and strengthened. But the next collapse, the Second World War, with its promotion of destructive technologies and development work instead of fundamental research, was to change the very nature of R&D at the intrafirm as well as at the national level.

Science and Industry. Modernisation Strategies in Norway 1900–1940[1]

FRANCIS SEJERSTED

1. *The Case of Norsk Hydro*

Although the establishment of Norway's largest industrial group, Norsk Hydro, has something of a fairy-tale aura about it, it also provides an excellent example of industrial activity genuinely based on science[2]. Norsk Hydro's foundation was based on the first commercially viable nitrogen fixation method for the production of fertilizer. The need for such fertilizers had been generally recognised, and methods of production were being sought in many laboratories in Europe and America. The Hydro method was based on generating a very high temperature by means of an electric arc, a principle developed by Professor Kristian Birkeland at the University of Oslo. After studying under Poincaré in Paris and Hertz in Bonn in the 1890s, he had taken up the study of the Northern Lights. This involved heavy expenditure on travel, the establishment of observation stations and the like. To raise money, he decided to put his knowledge of electromagnetism to industrial uses, and began, among other things, to develop an electric cannon. It was a flop, but during the work he noticed that in a strong magnetic field the electric arc was deviated into a flat disc perpendicular to the magnetic curves.

---

[1] The author's preparation of this final version has benefited from comments by participants in the ESF's Workshop. Other valuable comments have been received from Robert Marc Friedman, Kjetil A. Jacobsen and Sissel Myklebust. Parts of the article are particularly indebted to John Peter Collett, Videnskap og politikk. Samarbeid og konflikt om forskning for industriformål 1917–1930 (University of Oslo MA thesis in History, 1983).

[2] On the history of Norsk Hydro, see K. A. Olsen, Norsk Hydro gjennom 50 år. Et eventyr fra realitetenes verden, Oslo 1955; O. Gasslander, Bank och industriellt genombrott. Stockholm Enskilda Bank kring Sekelskiftet 1900, I-II, Stockholm 1956–59.

Shortly afterwards, Birkeland met Samuel Eyde, an engineer who had for some time been considering how best to exploit Norway's huge hydropower resources. Fertilizer production was one of the possibilities that had occurred to him. Eyde had been trained as a railway engineer at the technical college at Charlottenburg in Berlin. With his own engineering firm and major assignments behind him in Germany and Scandinavia, Eyde was a big contractor, with contacts in the financial world, especially in Sweden. At his chance meeting with Birkeland he happened to mention the possibility of finding a way of oxidising the nitrogen in the air by means of discharges of electricity. Birkeland saw that his electric arc might provide the necessary means.

The development of the method, and the building of the large installations required, became possible when the Swedish financier Marcus Wallenberg took an interest in the project. What apparently decided him was in fact that the Swedish electrotechnical group ASEA, in which he was heavily involved, was in difficulties. The large-scale development of Norwegian hydropower would open up a market for ASEA's generators and other products. Through Wallenberg's mediation, Norsk Hydro was founded on foreign capital, comprising both equity and debt. Eyde was made managing director of the new company, ASEA sold its products, and Birkeland got money to continue his research on the 'aurora borealis'.

2. Public Organization of Science for Industrial Purposes

The establishment of Norsk Hydro became a symbol of the potential for the exploitation of science. The decades leading up to World War I saw a general advancement of applied science in Norwegian society. The new "cult of science" manifested itself in all industrial countries, but was particularly noticeable in a country like Norway, where practical science had been virtually ignored before. Science, which in our context means natural science, now came to be regarded as one of the most important instruments for the modernisation of society. This belief was given particularly clear expression in 1892 by Professor W. C. Brøgger, one of the major builders of Norwegian scientific institutions:

Science stands for development and progress both spiritually and materially; all truly scientific research advances humanity, towards higher aims and in the last instance also towards greater happiness and improved living conditions[3].

[3] Cf. W. C. Brøgger, Verdens Universiteter I-III, Verdens Gang nos. 87, 88 and 92, 1892. Quoted from J. P. Collett, p. 23.

The rise to prominence of an ideology of science resulted, among other things, in the establishment of a number of public institutions for applied research and for research-based education, the most important of which were the Norwegian Agricultural University, the Fisheries Administration, and the Norwegian Institute of Technology, established in 1897, 1900 and 1911 respectively[4].

It is worth noting that the earliest of these institutions were devoted to the primary industries. The fact that the school of agriculture came first reflects the strong position, politically even more than economically, of agriculture. The establishment of applied science in that sector marks a natural conclusion to the huge expansion of public agricultural reform measures that had begun in 1874, a period during which public agricultural budgets increased tenfold[5]. There is also reason to believe that Norwegian agricultural research was among the international leaders, and the same can also be said of fisheries research. Roll-Hansen maintains that Norwegian fisheries research was the world's best during the decade before World War I[6].

Typically, too, one of the most successful projects in Norway in the 1900 to 1930 period was linked to the primary industries, namely the pioneering work of Professor Vilhelm Bjerknes in the development of modern meteorology. When it came to public funding of his project, it was the needs of agriculture and fisheries that decided the issue[7]. That large, practically-oriented scientific project became one of the greatest successes in the history of Norwegian science.

An interesting feature of the Norwegian situation is the relatively slow and hesitant development of organized public industrial research. Whereas our neighbours, Denmark and Sweden, established their polytechnic institutions in the late 1820s with the express aim of stimulating industrial development, Norway only got its first technical college in 1910![8] Scarcely

---

[4] Cf. N. Roll-Hansen, Johan Hjort og motstanden mot de praktisk-vitenskapelige fiskeriundersøkelser, in: N. Roll-Hansen, (ed.), Skandinavisk naturvitenskap og teknologi omkring år 1900 (Seminar report, the Institute for Studies in Research and Higher Education, the Norwegian Research Council for Science and the Humanities 1980:4), p. 89.

[5] Cf. K. D. Jacobsen, Teknisk hjelp og politisk struktur. En studie i norsk landbruksforvaltning 1874–1899, Oslo 1964, p. 9.

[6] Cf. Roll-Hansen, p. 91.

[7] Cf. R. M. Friedman, Appropriating the Weather. Vilhelm Bjerknes and the Construction of Modern Meteorology, Ithaca and London 1989. See especially pp. 208, 227 and 240.

[8] Cf. T. J. Hanisch/ E. Lange, Vitenskap for Industrien. NTH – En høyskole i utvikling gjennom 75 år, Oslo 1985, especially p. 14.

any other European country was as slow to establish a similar institution. The first proposal to create an institution for higher technical education was submitted to the Storting, the legislative assembly, in 1833, prompted by the establishment of a corresponding institution in Denmark. A group of professors at the University of Oslo were among the proponents of a plan which it would take nearly 80 years to realise. There are many reasons for this, including the distribution of political power: strong agricultural and regional interests were hostile to public expenditure in general and to industrial and urban ventures in particular[9]. There was also the predominance of small enterprises in Norway's industrial structure, with neither much need nor much ability to profit from research[10], and consequently little incentive to promote it.

The establishment of the Institute of Technology in 1910 gave rise to an intense discussion of the relationship between theory and practice. The tendency in established industrial and engineering circles was to emphasise practical training, and to insist that appointees to chairs must have had practical industrial experience. What actually happened was the contrary, based instead on ideas imported from Germany, where the modern view was that advanced technology ought to be regarded as applied natural science. So the professors who were appointed were primarily theoreticians, and the curriculum placed great emphasis on fundamental theoretical subjects: mathematics, mechanics, physics and chemistry. Some concessions were nevertheless made to practice: drawing offices and laboratory practical sessions figured prominently on the timetable[11].

In the laboratories, theory and practice would unite and ideas would be systematically tested. Laboratories were a major feature in the planning of the Institute of Technology, and not only for teaching purposes. They were intended to serve industry and to lay the foundations, through advanced new techniques and processes, for its further development. A thermal power and a hydropower laboratory were among the first to be completed, in 1912 and 1917 respectively, and both maintained close contacts with industry. Through its experimental turbine developments and research into construction problems, the hydropower laboratory, in particular, played an important part in the development of hydro-electricity in the interwar years.

---

9 Cf. ibid., p. 34.

10 Cf. F. Sejersted, A theory of Economic and Technological Development in Norway in the Nineteenth Century, in: Scandinavian Economic History Review 1, 1992.

11 Cf. Hanisch/Lange, p. 55.

The Institute also built other laboratories during this period, but if we are to believe a report from 1919, it was only in the two just mentioned that research for industry was carried out. The absence of industrial research at the electrical engineering, physics and chemistry laboratories[12] is all the more remarkable in view of the general importance of their subjects and their special significane to Norway. Limited capacity and the need to concentrate on teaching was one reason, but there was also the difficulty that the Institute was located in Trondheim, while industrial enterprises were mainly concentrated around Oslo.

There were, of course, contacts of a less formal nature, with Institute staff being employed as industrial consultants. This also seems to have been quite common at the University of Oslo. The physicist Birkeland has already been mentioned, in connection with Norsk Hydro, but he was only one of many. Companies also set up laboratories of their own, although often on a very modest scale. These less conspicuous trends were clearly quite widespread, but have yet to be properly charted.

3. Two Lines of Development

One feature of the developments around 1875 often referred to as "the second industrial revolution" was the growing reliance of technology on science. In Nathan Rosenberg's words, western industrial technology began to move away "from the visible world of levers, gears, cams, shafts, pulleys and cranks to the invisible world of atoms, molecules, electron flows, electromagnetic waves, inductance, capacitance, magnetism, amperes, volts, bacteria, viruses and genes"[13]. The source of renewal or innovation would be found at the point of contact between basic research in its own, relatively autonomous sector, and applied, production-related research in industry. This hook-up between science and technology is complicated, and a number of theories (not to mention ideologies) have been advanced concerning its nature.

There was of course a close connection between the new scientifically-based technology and the scientific ideology or "cult of science" referred to above. This ideology included a concept which has been called "the linear model", according to which real advances originate in independent

[12] Cf. ibid., p. 74 f.

[13] N. Rosenberg/L. E. Birdzell, Jr. (eds.), How the West Grew Rich. The Economic Transformation of the Industrial World, New York 1986, p. 252.

basic research. The findings of basic research can then be put into practice by means of applied research, development and production. It is the basic researcher, i. e. the pure scientist, who is the source of innovation and the real initiator, indeed who is the real moderniser. This view can be seen in the quotation from Brøgger above. The concept has lived on, with a varying degree of influence on research policy. In the period with which we are concerned, from the turn of the century on, it had considerable influence; it rose to prominence again after World War II, not least as a result of Vannevar Bush's famous report to the President of the United States[14].

Today the linear model, with its autonomous natural scientist as the source of innovation and the only true moderniser, is generally held to be out of touch with reality[15]. Even in those cases where there is a direct link between basic research and the process of innovation, as in the two major successes achieved by Professors Birkeland and Bjerknes, the process was much more complicated. Rosenberg argues that it became possible to redefine basic research in relation to this link between science and technology: "Basic science came to be seen as the testing and development of the explanatory structure of the natural sciences"[16]. It was often the case that ideas or processes that had been developed in industry could be improved or further developed on the basis of the basic researcher's scientific elucidation of the process. The "invisibility" of phenomena mentioned by Rosenberg as a characteristic of the new technology made it important for the industrial sector to have access to basic research expertise.

It can be argued that the industrial laboratory is a strategically important site which links science and technology[17]. The nature of such laboratories

---

[14] Cf. V. Bush, Science, the Endless Frontier, Washington 1945. Cf. E. Mansfield, The Economics of Technological Change, New York 1968, p. 169.

[15] Cf. N. Rosenberg, How exogenous is science?, in: ibid., Inside the Black Box. Technology and Economics, Cambridge 1982.

[16] Rosenberg/Birdzell, p. 253.

[17] Cf. ibid., p. 260. Taking the industrial laboratory as a gateway to the study of the link between science and technology is also strongly recommended by so-called "social constructivists" with close links to the sociology of science. Cf. W. E. Bijker/T. P. Hughes/T. Pinch (eds.), The Social Construction of Technological Systems, Cambridge 1987. There is also a school of thought which is more fundamentally opposed to the ideology of science and the linear model, holding that the key to the understanding of innovation lies in the study of work processes. From this point of view, no distinction is made between intellectual and manual labour. Technical change is associated with the experience in technical trades of work processes, organization and management. Cf. T. Halvorsen, Tekniske yrker og organisasjon. Ein kommentar til teoritradisjonar, Bergen, AHS series B 1991, no. 4.

has not, however, been unambiguously defined. They vary considerably in size, organization and function. Some are closely linked to production and engaged in routine testing, while others may be relatively autonomous and engaged in pure science, and most are somewhere between the two extremes. They may have been organized as formally independent units by private interests, branches of industry, or the authorities (e. g. the institutes at the Norwegian Institute of Technology). Most, however, were (and are) in the industrial sector. The more autonomous and scientific of the institutes, those which concentrated on the new, scientifically-based technology[18], tended to be founded at technical colleges and in large new industrial corporations. Norway had few if any big corporations; but it may be just as rewarding to focus on the emergence of the many small laboratories.

Industrial laboratories for systematic testing and for scientific experiments go a long way back in industrial history. It does seem, however, that the new technological paradigm of the second industrial revolution provided a particularly strong motivation for the establishment of laboratories. Rosenberg refers to a study showing that there were 139 laboratories in American industry in 1898, whereas a further 553 had been added by 1918[19]. There are clear indications that the number of industrial laboratories also increased in Norway around the turn of the century.

There was thus a multiplicity of forms of organization at the interface between science and technology, but among them two main lines or strategies can be made out, based on two fundamentally different views of the nature of that interrelation. The first reflects "the cult of science" and the linear model we have already considered. Its ideology of science was to influence public research policy, but not so as to place the emphasis wholly on pure basic research. The idea was rather that modern industry ought to be based on the relatively independent generation of knowledge, with practical scientists deciding the agenda. Norsk Hydro was a prominent example of a particular industry based on independently developed knowledge. In effect, the Institute of Technology became an embodiment of the same view. One characteristic feature of this strategy was that research laboratories tended to be located at a distance from production sites.

The alternative view held that research should, to a great extent, be based on industrial premises or be designed to meet production needs.

---

[18] Cf. A. D. Chandler, Scale and Scope. The Dynamics of Industrial Capitalism, Cambridge/Mass. 1990, pp. 32 ff.

[19] Cf. Rosenberg/Birdzell, p. 247.

The main objective in this case was not the production of fundamental knowledge, but the potential of research to develop existing technology still further. Many industrial laboratories were engaged in routine testing pure and simple, much of which can hardly be called "research". But one can still see this as a success for the scientific ideal, since problems were being approached systematically and objectively. It was a matter of applying the methods and procedures of science to problems defined by others, outside of the laboratory, primarily those engaged in the ongoing production process.

It is important to note that internal industrial laboratories of this kind were often also intended to build up receptiveness to new ideas and knowledge, and the competence to make use of them. Such input might have many sources, including the scientific community. A good illustration is provided by the contributions of basic research to the understanding of "invisible" processes, which we have touched on above. Whereas, according to the linear model, science of the more autonomous kind was the source of technological innovation, in this view it is seen as one among many aids in a more complex innovative process. The two diverging concepts of the interrelation between science and technology have implications for the shape given to the links in practice. One strategy is firmly based on scientific premises, and the other on the premises of technology or existing production. The possibility that both can bear fruit cannot be ruled out.

Both historians of technology[20] and economists who analyse technological change[21] have adopted concepts such as "technological systems" or "trajectories", and distinguish between various forms of "radical" innovation on the one hand and "conservative" or "incremental" innovation on the other. Truly epoch-making or "revolutionary" innovations change systems, whereas "conservative" ones improve existing concepts and tend to reinforce systems. Between them (in Freeman's taxonomy) we have "radical innovations", which can be generated not least in relatively autonomous scientific laboratories. If technological systems can be thought of as having life cycles, radical innovation is most likely to occur during their initial phases.

The second industrial revolution, with its new scientifically-based technology, was a change of the technological system. Indeed it might be more

---

[20] See for instance T. P. Hughes, The Evolution of Large Technological Systems, in: Bijker/Hughes/Pinch (eds.), Social Construction, p. 51–82.

[21] See for instance C. Freeman/C. Perez, Structural crises of adjustment: business cycles and investment behaviour, in: G. Dosi/C. Freeman/R. Nelson/G. Silverberg/L. Soete (eds.), Technical Change and Economic Theory, London 1988.

appropriate to speak of a number of new systems or "trajectories". Because it had an industrial structure consisting of a large number of small units, and lacked institutions for applied science, Norway was poorly equipped either to exploit, or to contribute to the development of the new technological systems. On the other hand, the country did have hydropower: with electricity as the major cornerstone of the new systems, at least one important prerequisite was in place. And Norway was quick to develop new science-based electrochemical and electrometallurgical industry. But it was not born of a strong scientific or technological environment: the impulses came from outside. Norsk Hydro was an exception, in that the very radical innovation on which it was based was Norwegian.

What Norway needed were strategies for meeting the challenges presented by the new science-based technology. Among the steps taken by the authorities, as we have seen, was the establishment of scientifically-based higher education and of laboratories for applied science. Results were achieved, particularly in the primary industries and to some extent also in the manufacturing sector. Even at this early stage, examples can be found of relatively radical industrial innovation[22]. This strategy was influenced by the "cult of science" of the day, and we will follow its further development through the World War I period. A second strategy, though perhaps less spectacular and less coordinated, was to prove important in the long run: it consisted of expanding production-related research and development, and was intended to develop sufficient competence to exploit knowledge from abroad instead of producing it at home. We shall return to this second strategy in the last section.

4. *Developments during the World War*

The First World War was a powerful stimulus in putting science to practical uses. In Britain, a separate Department for Scientific and Industrial

[22] One example comes from the chemical pulp industry: At Kellner-Partington's Norwegian Borregaard plant, Kellner's sulphite process was further developed by the chemical engineer Nils Pedersen on the basis of a scientific explanation of the process. That enabled Borregaard to supply a quality product which sustained it through crises. Cf. T. Bergh/ E. Lange, Foredlet Virke, Historien om Borregaard 1889–1989, Oslo 1989, p. 33 f.; and S. Samuelsen, Papirindustrien og forskningen, in: Industri og forskning (Commemorative volume in honour of Johan Throne Holst), Oslo 1938, p. 119. A second example is Carl Fr. Holmboe's pioneering contribution to the development of originally German oil-hardening patents, which laid the foundations for De-No-Fa A/S. Cf. C. F. Holmboe, En norsk innsats i forskning og teknikk. Trekk fra oljeherdingens historie, Oslo 1947, p. 24 f.

Research was established, and in the USA the National Research Council was formed at the behest of the National Academy of Sciences. The latter in particular was held up as an example for Norway to follow. Although Norway was not among the belligerents, her industry soon ran into difficulties, chiefly because of raw material shortages. That was the position when, on behalf of "Norwegian science", Professor W. C Brøgger made an offer of cooperation with the public authorities, which they accepted. A Ministry of Industrial Supplies and a Raw Materials Committee were set up, through which funds were channelled to industrial research. That research, prompted by the wartime lack of raw materials, was to produce lasting results in some areas[23].

The main driving force in the Raw Materials Committee was William Goldschmidt, Professor of Geology at the University of Oslo. Brilliant at pure research, he also fully appreciated the value of applied science. From 1916 on, he also held an appointment as scientific consultant to Elektrokemisk A/S (Elkem). This brought the research interests of the authorities, the universities and industry together, with interesting results.

Elkem had been established in 1904 to put Norwegian hydropower to industrial use. The prime mover behind its establishment was Sam Eyde, and the first item on its agenda concerned nitrogen. When that activity had reached a certain level, in 1905, it was a natural move to hive it off into a separate company, Norsk Hydro. This was to become Elkem's usual procedure. The company defined itself as an innovator, whose aim it was to get new enterprises established in the power-intensive sector. It built up a research environment in electrochemistry and electrometallurgy[24]. In financial matters, as well, Elkem was an innovator, operating in practice as Norway's first venture capital company by selling its shares in enterprises once they began making operating profits[25]. Elkem for a time functioned successfully as a research-based innovative company. It was, for instance, instrumental in the establishment of the Norwegian aluminium industry when it took the initiative to establish *Det Norske Nitridaktieselskab* in 1910. (The underlying process, however, was based on a licence from abroad.) What was taking place was the assembling of the components for the construction of a new research-based industrial sector.

---

23 Cf. J. P. Collett, p. 12. Collett's work as a whole presents a searching analysis of the many public efforts to promote Norwegian research in the interests of industry.

24 Cf. E. Petersen (ed.), Elektrokemisk A/S 1904–1954, Oslo 1953, pp. 50 ff.

25 Cf. S. Knutsen, Bank, samfunn og økonomisk vekst (University of Oslo M. A. thesis in History, 1990), pp. 138 ff.

The situation during World War I of an economic boom, together with the technical challenges presented by the shortage of raw materials, appears to have benefited Elkem. Goldschmidt was appointed and a number of applied science research projects were initiated[26]. Among other things, Goldschmidt worked out methods for producing aluminium from Norwegian labradorite. Production was started and transferred to Norsk Hydro, but became unprofitable when normal trade relations were resumed after the war. However, two of the research projects undertaken during the conflict were to prove highly significant. One concerned the manufacture of titanium dioxide pigments, and the other was the Söderberg electrode.

The development of the sulphuric acid process for producing titanium dioxide pigments originated in pre-war attempts to remove titanium oxide from iron ore. It was discovered by chance that the titanium compounds had excellent covering properties. The further development of the process involved the highest expertise and bore some of the hallmarks of pure science[27]. Goldschmidt, for instance, contributed a theoretical study, based on the latest crystal-optical methods, of the physical properties of pigments[28]. Production of titanium dioxide pigments on an industrial scale began in 1916, and was transferred to the newly-established company Titan Co. A/S. In 1918, production of titanium dioxide pigments was begun in the USA, but the Norwegian sulphuric acid process proved superior, and in 1920 the Americans bought the Norwegian rights from Titan Co[29].

The point of departure for the Söderberg electrode was the wartime scarcity of electrodes for the electrical smelting of metal. In 1916, Elkem launched a project aimed at developing electrodes that could be manufactured in Norway. From the 1870s on, electrical smelting furnaces underwent very rapid development into ever-bigger units. Electrodes had become the bottleneck limiting further development; this was a question not only of size but also of having to interrupt the smelting process whenever an electrode had to be replaced. The problem was solved by the continuous self-baking electrode developed by the Elkem engineers Carl W. Söderberg, Mathias Sem and Jens Westly.

In 1922 Elkem successfully adapted the electrode for use in the production of aluminium by electrolysis. It also became the central element

---

[26] Cf. Petersen, pp. 66 ff.

[27] Cf. ibid., p. 78.

[28] Cf. Collett, p. 141.

[29] Cf. D. A. Hounshell/J. K. Smith, Science and Corporate Strategy. Du Pont R&D, 1920–1980, Cambridge 1988, pp. 211 ff.

in the development of the Tysland-Hole furnace for the electrical smelting of pig iron, a project with which many, and not least Elkem, had been concerned for several years. The first Tysland-Hole furnace was put into operation at Christiania Spigerverk in 1925. The Söderberg electrode is one of the most successful research-based innovations in the history of Norwegian industry. Rights to use it were sold all over the world and were for many years Elkem's main source of income[30].

5. Setback for the Grand Strategy

In many respects, the cooperation established during World War I between science, government and industry with a view to solving the raw materials problem was a success. It also prompted the idea of corresponding arrangements of a more permanent kind. A "Central Committee for scientific cooperation for the advancement of industry" was established in 1919, to extend the concept of scientifically-based and scientifically-controlled industry into peacetime conditions. It became the forum for much of the country's debate on research policy[31]. In the same year, the Storting set up a three million krone research fund, and A/S Norsk Varekrigsforsikring set up another fund with a similar amount[32].

Relations between the Ministry of Industrial Supplies and the Central Committee were to be marked by a divergence between the two outlined above. Majority opinion in the Central Committee favoured industrial research on scientific premises, but the plans which emerged in due course for six research institutes bore the stamp of the Ministry's wish for production-related laboratories[33]. Only two of them were established, however, for the wood-processing and canning industries (see below). They received public financial support, as did the development under Elkem auspices of the Tysland-Hole furnace (see above). In general, though, the coordinated public and private efforts to promote scientifically-based industry only produced meagre results. There would be no breakthrough for government funding of industrial research until after World War II. Then, too, the grand strategy dominated, with science expected to lay

---

[30] Cf. G. Nerheim, Fra teknologiforskningens barndom i Norge, oppfinnelsen og utnyttelsen av Söderberg-elektroden, in: N. Roll-Hansen (ed.), pp. 19–32.

[31] Cf. Collett, pp. 103 ff.

[32] Cf. ibid., p. 188.

[33] Cf. ibid., pp. 119. 131 and 137.

new foundations for new industries. This found its clearest expression in the generous funding of nuclear research[34].

The direct cause of the public sector's withdrawal in the interwar years, and accordingly of the absence of one of the preconditions for realising anything approaching the grand strategy, was the economic slump in 1920–21, which tightened industrial and government budgets alike. As Collett writes, central government lent a hand when raw materials were scarce, but not when markets failed[35]. It was of course easier for the authorities to step in under wartime conditions, which could be described as exceptional. In peacetime, industry should be capable of going it alone – the manufacturing sector, that is, but not the primary industries. Unfortunately, however, industry lacked both the necessary organization and the relevant traditions, and proved incapable of generating new research-based activities on its own. Under pressure, it backed out.

As has been mentioned above, both the wood-processing and the canning industry did acquire research institutes, in 1923 and 1931 respectively. Both industries were marked by a relatively large proportion of small units, mutually dependent on one another's quality when launching their products in foreign markets, and with plenty of room for minor improvements in their manufacturing processes. Their research was close to the production line, and much of it was related to quality control[36]. That is not to say that the institutes were not important to their industries. The wood-processing industry in particular appears to show that a relatively large institute capable of keeping its distance from actual production can generate research of its own which proves to be of great practical value to the industry, although its original purpose was to solve immediate production problems. One of the more theoretical and long-term problems the institute worked on concerned the bleaching of wood pulp. In connection with this, a photoelectric meter for measuring whiteness was designed which became a worldwide success. It was, however, sold to Zeiss since Norway lacked the industrial resources for its manufacture[37].

We have seen independently generated research aimed at industrial development being carried out in certain areas and within certain major industries. One case was the hydropower laboratory in Trondheim. Elkem

[34] Cf. O. V. Nilsen, Noratom og drømmen om en norsk atomindustri (TMV working paper no. 54, 1992).

[35] Cf. ibid., p. 262.

[36] Cf. G. Lunde, Forskning i hermetikkindustrien ; Samuelsen, in: Industri og forskning and Collett, pp. 233–238.

[37] Cf. Samuelsen, p. 121.

continued its industrial research, but on a very reduced scale after the recession[38]. A certain amount of research and experiment of a more fundamental nature was being carried out, however, in the metallurgical industry. The "Orkla process" for the production of pure sulphur from the pyrite from the Løkken works was one example of an important process developed in Norway[39]. But what had become of Norsk Hydro, the giant in Norwegian scientifically-based industry?

It became clear relatively soon that Hydro's nitrogen fixing method would run into competition. In Germany a major development during the war was based on the Haber-Bosch process for synthesising ammonia, and when peace came the Germans were seen to have a big lead in this area[40]. Hydro responded with an offensive strategy, completing a large research laboratory in 1919. To head the laboratory they appointed B. F. Halvorsen, Professor of technical inorganic chemistry at the Institute of Technology. But the investment was in vain: by 1927 Hydro had to admit defeat and enter into an agreement with IG Farben, the holders of the Haber-Bosch patent. IG Farben acquired a 25% stake in Hydro, which was to carry on production according to IG's patent and sell through its marketing organization. A nitrogen cartel called DEN (Deutschland-England-Norwegen) was formed in 1930 which divided up the world market and proved highly efficient[41]. The maintenance and further development of its scientific and technological lead was crucial to IG Farben's policy of building and dominating cartels[42]. A natural consequence of the integration of Hydro into the system was the closure of its large research laboratory. Research was limited to that which was relevant to production and moved to production sites.

Hydro's fate was not exceptional. In the 1920s, Det Norske Zinkkompani undertook the development of its own electrothermic zinc process, but was obliged in 1929 to buy an electrolytic process developed in the USA. Similarly, independent Norwegian nickel production had been taken over by the Canadian company Falconbridge, which had its own technology[43].

---

[38] Cf. Petersen, pp. 99–108.

[39] Cf. Collett, p. 243.

[40] Cf. A. Olsen, p. 286.

[41] Cf. H. G. Schröter, Aussenpolitik und Wirtschaftsinteresse. Skandinavien im Außenwirschaftlichen Kalkül Deutschlands und Großbritanniens 1918–1939, Frankfurt 1983, pp. 265–68.

[42] Cf. ibid., pp. 248 and 252 ff.

[43] Cf. Collett, p. 243. See also G. C. Wasberg/A. Strømme Svendsen, Industriens historie i Norge, Oslo 1969, pp. 185–87.

The development of hydropower might have been expected to create a favourable climate for electrotechnical suppliers. Norway was a rapidly growing market for almost any product from that industry. Investments in the industry's generating machinery accordingly rose steadily and rapidly from 1914 to 1930[44]. This was an industry which was at the leading edge of the technological development of the time and was to a large extend founded on research. It has indeed been argued that although Norwegian production in this area was initially based on foreign licences, it soon had a stimulating effect in Norwegian technological circles[45]. Sissel Myklebust, on the other hand, puts the opposite case[46]. She points out that, with regard to both low-tension and high-tension products, this branch of industry remained underdeveloped in Norway. Whereas the industry recorded export surpluses in most other European countries, developments in Norway were marked throughout by large import surpluses. As we saw, Wallenberg's involvement with Norsk Hydro was motivated by the prospect of delivery contracts for the Swedish company ASEA; and that was exactly the way things turned out. Generally speaking, Norway became a market for foreign suppliers, and such manufacturing as was undertaken in Norway was, with rare exceptions, under licence. Given the natural conditions, it was only to be expected that Norway would develop an electricity-based and power-intensive industry. It was also research-intensive, and we have seen that science, industry and government all had, to begin with at least, high hopes of developing the independent production of knowledge in connection with it.

The system comprised two elements: the power-intensive industry itself, i. e. the electrochemical and electrometallurgical industries, and its suppliers, chiefly the electrotechnical industry. Power-intensive industry, which had to be located close to the source of energy, naturally expanded rapidly in Norway. Inspired by the spectacular Norsk Hydro success, attempts were made to develop domestic production of knowledge on which to base further development; but that strategy failed. It can clearly be seen how the growth of foreign ownership between the wars parallels the abandonment of risky knowledge-production in favour of increasing reliance on foreign licences[47]. Where electrotechnical manufacturers are con-

---

[44] Cf. Wasberg/Svendsen, p. 183.

[45] Cf. T. Bergh/T. J. Hanisch/E. Lange/H. Pharo, Norge fra u-land til i-land. Vekst og utviklingslinjer 1830–1980, Oslo 1983, p. 162.

[46] Cf. S. Myklebust, 'Den nye arbeidsdagen'. Det 20. århundres norske industrihistorie (TMV working paper no. 41). See especially pp. 18 ff.

[47] Cf. A. Stonehill, Foreign Ownership in Norwegian Enterprises, Oslo 1965.

cerned, on the other hand, geographical location was no constraint. Norway failed to avail itself of the opportunities presented by the large domestic market. Advanced equipment was, to a large extent, imported. This strengthened the impression given by electricity-based heavy industry as a whole of being an enclave[48].

It is also worth noting that even in those cases where new knowledge was developed initially, this was not followed up by the consolidation needed to retain the technological or scientific lead thus gained. In short, it proved to be easier to buy into new technology than to develop it independently.

There were, nevertheless, certain sectors of the mechanical engineering industry which did expand as suppliers to power-intensive industries and did manage to build on leading-edge competence. Turbine manufacturers are a striking example. But heavy mechanical engineering, which in Norway means primarily shipbuilding, gives the same impression as the electrotechnical industry. Developments in this field served to highlight the unfavourable structural conditions for a knowledge-producing industry.

The transition from sail to steam in shipping was closely related to developments in Norwegian shipbuilding. Before the First World War, Norway's shipbuilding industry was the biggest in the Nordic countries, but after the recession in 1920 it collapsed almost completely. That meant that it was not alive to the transition from steam to motor vessels. In the interwar years, Norwegian shipowners largely ordered their new vessels from foreign yards, leaving Norwegian yards to fall back on repairs.

The reasons for this development have been debated[49]. In part, they were economic: suppliers would not extend sufficient credit, and the back-to-par policy for the currency had damaging effects. But some of the reasons must also clearly be sought in the inability of Norwegian shipyards to meet the technical challenge posed by the transition to motor vessels.

---

[48] Myklebust discusses the enclave features of the industry, also heavily emphasised in: Å. Cappelen/J. Fagerberg/L. Mjøset, Den norske modellen, in: Nordisk Tidsskrift för Politisk Ekonomi 15–16, 1984. It is important to note, concerning the electricity system, that it consisted of two parts. One was the enclave-like system just described, comprising export-competing heavy industry, with the state and private foreign interests developing the power supply. The other was the general power supply system developed by local government and supplying small industries, thus giving them scope for technological modernisation. Cf. L. Thue, Statkrafts historie, (unpublished Ms).

[49] Cf. H. W. Andersen, Norsk skipsbyggingsindustri gjennom 100 år, in: E. Lange (ed.), Teknologi i virksomhet. Verkstedindustrien i Norge etter 1840, Oslo 1989, pp. 89 ff.

According to articles in the 1924 volume of the periodical Skibsbygning, there was a shortage not only of engineers with the appropriate qualifications, but also of owners able and willing to make the heavy investments needed to begin manufacturing diesel engines[50].

One shipyard, Akers mek, eventually began manufacturing diesel engines under licence from Denmark's Burmeister og Wain, who were pioneers in the field. Symptomatically, small whaleboats continued to be built at Norwegian yards: there was a domestic market, and the boats were fitted with steam engines because it was thought that the noise of diesel engines frightened whales[51]. It has been said, with considerable justification, that "the technological gap when the transition came from steam to motor vessels" contributed to "the economic catastrophe in Norwegian shipbuilding between the wars"[52].

Most Norwegian yards were too small to take the strain of introducing modern, scientifically-based technology. A corresponding structural inadequacy stands out especially clearly in connection with the production of engines for the large Norwegian fishing fleet. In the interwar period, most fishing boats had been fitted with very simple hot-bulb engines, which Norwegian engine manufacturers had been quick to start producing at the turn of the century, when the fishing fleet was motorised. Between the wars, however, the diesel engine had been developed into a much better alternative. Professor R. Lutz had conducted laboratory experiments with small diesel engines at the Institute of Technology, and strongly urged Norwegian manufacturers to make the transition, in cooperation with the Institute. But industry responsed negatively to the professor's initiative[53]. It was not prepared to be guided by scientists.

In general, the time was past when scientists, or basic researchers, could take the lead in modernisation. For them ensued a period of lost opportunities and frustration. As we have seen, they did achieve breakthroughs in certain areas, for instance to some extent in the primary industries. One of the best examples of basic research aimed at practical results was Bjerknes's big project, discussed above, which laid the foundations for modern meteorology. It is all the more striking, then, to find him giving clear expression to the general frustration; in 1938 he wrote that "we lack the tradition, the environment and the organization for providing talented

[50] Cf. Skibsbygning 2, 1924, pp. 1, 20, 36 and 128 ff.

[51] Cf. Andersen, p. 93.

[52] Cf. Hanisch/Lange, p. 100.

[53] Cf. ibid., pp. 120–24.

individual research scholars with opportunities, means, and a route to utilization"[54].

The reasons for the failure of the grand strategy of creating independent competence-building in the electrochemical and electrometallurgical industries are partly that the structure of Norwegian industry did not lend itself to such a strategy, and partly that the scientific environment was not vigorous enough. The fragility of this platform is confirmed by what happened in the mechanical engineering industry. Things might have turned out differently but for an economic crisis and withdrawal by the authorities.

Reasons for the overall failure of the grand strategy despite a number of distinct successes must, however, also be sought in the weaknesses in the concept itself. The linear model was an over-simplification. Although in practice the complications of reality were also taken into account, generally speaking the means for making the links between science and technology work were poorly understood. It could also be argued, in line with Freeman's theories, that the period of more radical innovations in power-intensive industries was over. Under these circumstances, there may have been some sense in the less spectacular and less deliberately coordinated strategy of focussing on closely production-related R&D.

6. Rationalization and the Second View of Technology

Beginning in 1933, an economic recovery brought with it a steady rise in industrial employment. This was not due to new research-based industries, rather the contrary: the most rapid growth was achieved by small enterprises geared to domestic markets. Nevertheless, it did represent technological innovation, firstly inasmuch as one of its most important preconditions was the rapid expansion of the general electricity supply system, and secondly because it involved a high degree of innovation in product technology. This innovation appears to some extent to have been generated by the crisis[55]. Another striking feature is a sharp increase in the employment of engineers. In the 1920s, many of the graduates of the new Institute of Technology had had to emigrate. The Norwegian industrial system had no use for its new experts, who had no difficulty in finding

[54] V. Bjerknes, Norge i Forskningsalderen, in: Industri og Forskning, p. 187.

[55] Cf. note 48 above and see F. Sejersted (ed.), Vekst gjennom Krise. Studier i norsk teknologihistorie, Oslo 1982.

work abroad[56]. In the 1930s they came home. Many of them must have found jobs in production-related research laboratories.

So both research and researchers were being put to increasing use, although most often on projects determined by others. Research in the primary industries and in the branch research institutes of the canning and paper industries was mentioned earlier. Some of it was carried on in a grey area between the two strategies. The establishment of industrial laboratories in close connection with production must be seen as part of a larger project of modernisation. As Minister of Trade, Lars Oftedal, said at the inauguration of the canning industry's laboratory in 1931, "competition no longer takes place in factories, but in laboratories"[57].

As has already been mentioned, industrial laboratories were nothing new. But although they existed earlier, they only seem to have gained a real foothold in Norway with the second industrial revolution. The Alf Bjerke A/S paint factory presents a minor but sufficiently typical example from the early days. Established in 1880, it gradually became quite a sizeable operation. The company history tells us that after a few years, they "equipped a small laboratory" where "experiments" were carried out, and continues: "Looking through formulae from the factory's earliest days, one recognises a number of the standard recipes from the sparse paint and varnish literature of the time, and it is surprising how relatively quickly they were altered in accordance with Bjerke's own experience". Though the work in the laboratory may scarcely have merited the term research, it did consist of the systematic collection and testing of formulae from abroad. As we turn the pages to the 1930s, we reach a watershed: "From now on, the paint industry increasingly turns to real scientific research"[58].

In the meantime, the number of laboratories in Norwegian industry had grown steadily, but in the 1930s there are signs of acceleration. The pharmaceutical company Collett & Co A/S is a good example. It was established in 1933 to manufacture vitamins, one of the new science-based products of the period. For an enterprise of its kind, research was a matter of course: "Modern physical measuring instruments were bought for the laboratory, and appropriate methods of analysis were worked out for all vitamins. By means of analysis and experiment, the most favourable conditions were arrived at for preservation of the sensitive substances"[59]. This

---

56 Cf. Hanisch/Lange, p. 101.

57 G. Lunde, p. 105.

58 Firmaet Alf Bjerke gjennom 75 år, Oslo 1955, pp. 35–6 and 49.

59 Collett & Co A/S, Oslo 1958.

hardly amounts to pure research, but the laboratory does seem to have helped the company to build up the necessary competence for assimilating and exploiting fundamental new knowledge from outside sources.

An incomplete survey of industrial research was made in 1940. It was interrupted by the war, but gives certain indications. Questionnaires were sent to 200 enterprises, of which 126 replied. As many as 75% said they had or were in the process of acquiring laboratories of their own. Of those with laboratories, 30% replied that they carried out research. The largest numbers of laboratories were in the fields of ore and metal extraction, and in the chemical and electrochemical, food and beverage, and iron and steel and non-ferrous metal industries[60]. Norway clearly shared in the general spread of industrial laboratories. But only part of their work was pure science; research was integrated into a larger project.

The extent to which things had changed is illustrated by Johan Throne Holst, one of the period's captains of industry. Under his management the big Freia chocolate factory became a model. One of this major trend-setter's most important characteristics was precisely his commitment to research. He had been influenced by the "cult of science" during World War I. In the 1916 to 1919 period, he earmarked 200,000 kroner for a company laboratory with up-to-the-minute equipment for "scientific experiments in the food industry". At that time, money was plentiful; but he continued to follow up his interest in the interwar years, both by giving donations to the University and by expanding research relevant to his own business. The success of the company, he wrote, "depends on its ability to keep up with theoretical developments in its sector and to take the lead in its practical exploitation of theoretical advances"[61]. This was principally a case of science in the service of production on the factory-owner's terms. The aim was to be at the forefront, not of theoretical development, but of its practical exploitation. Nonetheless, Throne Holst's commitment reflected a strong general faith in science.

Throne Holst was also quick to see the potential for applying scientific methods to the rationalization of the production process. In other words, he extended the range of his interests beyond technology, to embrace

[60] Committee report on the organization of technological research, Oslo 1946, p. 16 f. Cf. H. Skoie/J. P. Collett, Teknisk-industriell forskningsorganisasjon i Norge 1945–80. Prinsipiell debatt og hovedlinjer i utviklingen. A report from the Institute for Studies in Research and Higher Education of the Norwegian Research Council for Science and the Humanities. Annex 3 to the report on public support for technical industrial research and development in Norway (Norwegian Official Reports 1981: 30B), p. 99.

[61] E. Rudeng, Sjokoladekongen, Oslo 1989, pp. 158 and 250.

sociological research as well. He dreamt of "a social order in which the commercial sector would also be built on rational foundations". One fruit of this commitment was his contribution to building up social research at the University of Oslo[62].

As a representative of interwar enthusiasm for rationalization, Throne Holst was fairly typical. His enthusiasm for practically oriented research was gradually incorporated in a broader concept of systematisation and rationalization. Trends towards "scientific management" can be seen in many countries between the wars. Its most vocal supporter in Norway was another leading industrialist, Joachim Lehmkuhl, whose book Rationel Arbeidsledelse (rational management) was published in 1920. It is concerned with a technocratic project, in which the engineer has taken over the leading modernising role, and the natural scientist has been moved into the wings. The engineer's function is to mediate between natural science and a positivistic social science, and to do so from a platform beyond either. The project could be seen as a victory for science: society is to be made scientific, but only at the expense of science's autonomy.

Although these technocratic ideas were quite widely held and greatly stimulated practical research, it has to be said that their realisation was not particularly conspicuous in Norway. Lehmkuhl himself arrived at the conclusion that the structure of Norwegian industry set clear limits on the implementation of his ideas. He built up a relatively successful electrotechnical company which manufactured some of the new durable consumer goods of the day. But it was not until he moved to America that he was really able to build up an enterprise for rational mass production of a durable consumer good, the Timex watch factory[63]. It has also been pointed out that Norway's industrial relations system restricted the scope of management to exercise techno-economic rationality. Norway had a distinctive link between job content and collective bargaining, a link not found for instance in Germany[64]. This is a feature of Norway's special brand of capitalism. In some instances, and perhaps especially where radical innovation was concerned, it could retard technical adaptation.

---

[62] Ibid., pp. 251 and 304.

[63] Information from Kjetil Ansgar Jakobsen, who is preparing a dissertation on technocratic trends in Norway in the inter war period. I am indebted to Jakobsen for several valuable comments on this part of the article.

[64] Cf. R. Sakslind/T. Halvorsen/O. Korsnes, Arbeidslivsforskning og de særegne industrielle relasjoner i Norge, in: I. Bleiklie/P. Bomann-Larsen/E. Falkum/T. Gran/T. Jensen (eds.), Politikkens Forvaltning (Commemorative volume in honour of Knut Dahl Jacobsen), Oslo 1985, pp. 58 ff.

Towards the end of our period, however, concepts of rationalization won a certain general acceptance.

As we have seen, the two strategies outlined here mirror two different views of the link between science and technology. On the one hand, technological development is linked to scientific discovery, with science seen as the source of technological innovation. On the other, technology takes centre stage and science, including pure science, is regarded as one aid among many. We have followed a certain degree of development, from a strategy based on the former view and towards a practice (rather than a strategy) based more on the latter, after the grand strategy had been abandoned. But we have also noted how that practice rested on a number of specific ideas which we have labelled "technocratic", implying a tendency to regard society as a whole as a technical system, something like a machine, in which technology (in the narrow sense), the social organization, and the individual must be adapted to each other so that everything functions efficiently and smoothly. A society of that nature presupposes government by experts, and what has often been called "the art of social engineering". Science would be an important aid, but the technologist is the leading actor. The point of departure, however, is in science. Scientific thinking is fundamental, not only to the social sciences, but to the government of society. Technocracy had a strong position as an ideology in the 1930s. How far it extended into practice is another matter.

Certainly industrial development was influenced by technocratic thinking, and one can see where Taylor made his mark. It makes sense to see the spread of industrial laboratories, too, as a manifestation of this way of thinking. Company laboratories whose immediate concern was production brought systematic procedures and rationality into industrial development, and were successful in introducing technological novelties from outside – at least if they were not too radical. The new awareness of the complex inter-relations between technical development and social structure was in some respects a better concept than the linear model which underpinned the grand strategy, but the technocratic concept of rationalization had weaknesses of its own.

The Evolution of European Industrial Technology in the Interwar Period

JOHN CANTWELL[1]

1. Introduction

A tradition has developed in which patent statistics are used to examine historical trends in the evolution of industrial technology, and the relationship between these trends and various other economic variables. This tradition goes back at least to Schmookler, who provided a more comprehensive treatment than had earlier studies[2]. The major deficiency of this literature as it has developed is its inadequate allowance for the structural characteristics of technological change. In one respect it is surprising that little attention has been given to the structural dimension, as opposed to the time series form of analysis. After all, patent statistics are ideally designed to provide a highly detailed disaggregation in various ways. The difficulty has been that until now the details required have not been collected in a computerised form, and carrying out such an exercise is a major task. In his study Schmookler spent a year at work in the Patent Office just to devise his classification scheme[3].

Three areas stand out in which the use of patent statistics could make a potentially strong contribution to current discussions, but on which the evidence has so far been lacking. The first is in making international

[1] The author gratefully acknowledges the support of the UK Economic and Social Research Council. He also wishes to thank Jane Myers and her staff at the US Patent and Trademark Office for their invaluable assistance, Pilar Barrera for her help in compiling the data used in the tables, and Peter Kenway for help with computer programming. He is grateful to François Caron, Mark Casson and David Edgerton for helpful comments on an earlier draft, and to the participants of the ESF workshop.

[2] Cf. J. Schmookler, Invention and Economic Growth, Cambridge/Mass. 1966.

[3] Cf. ibid., p. 21.

comparisons of technological change. National economies each have their fields of comparative advantage in technological development, and achieve different rates of innovation[4]. The countries that sustain the fastest rates of local innovation tend to be those whose comparative advantage lies in the fields of greatest technological opportunity. Consequently, it now seems to be quite widely accepted that major new sources of technological opportunity may be associated with changes in international leadership in innovation. The shift in dynamism from Britain to the USA and Germany historically, and the more recent shift from the USA to Japan were both linked to the rise of new fields of technology and related organisational methods. The relationship between changes in international leadership and changes in technological opportunities across activities or sectors is not as yet well understood.

Secondly, the overall rate of innovation depends upon the interchange between different fields of technological activity. In particular, innovation tends to occur most rapidly where a new source of technological opportunity has a positive impact on opportunities in related fields typically established earlier. This idea has gained currency in the long wave literature, in which it has been proposed that longer periods of innovative growth are associated with the consolidation of a new technological paradigm based on some characteristic kind of activity, with its most immediate and fullest range of applications in selected carrier branches which become the highest growth sectors[5]. However, this is at sharp variance with existing empirical studies of long waves, which tend to examine only the course of aggregate patenting or major inventions[6]. At this broad level of analysis the discussion seems to have reached something of a stalemate, and is unlikely to advance without paying closer attention to the structural changes which underpin longer term shifts in the overall rate of innovation.

Thirdly, just as countries and industries achieve technological progress at different rates, so do companies in the same industry. The growth of firms depends upon the specific capabilities that they have accumulated,

---

[4] Cf. J. Cantwell, Historical trends in international patterns of technological innovation, in: J. Foreman-Peck (ed.), New Perspectives on the Late Victorian Economy: Essays in Quantitative Economic History, 1860–1914, Cambridge 1991.

[5] Cf. C. Freeman/C. Perez, Structural crises of adjustment: business cycles and investment behaviour, in: G. Dosi, C. Freeman et al. (eds.), Technical Change and Economic Theory, London 1988.

[6] Cf., for example, S. Solomou, Innovation clusters and Kondratieff long waves in economic growth, in: Cambridge Journal of Economics 10, 1986, or A. Kleinknecht, Innovation Patterns in Crisis and Prosperity: Schumpeter's Long Cycle Reconsidered, London 1987.

which differentiate them from their competitors[7]. Within an industry each firm has a comparative advantage reflected in its technological position, or the fields in which its expertise is relatively strongest. To assess the competitiveness of firms, and to examine the specific path followed by the evolution of the technology developed by each of them, it is necessary to draw a clear distinction between sectors of technological activity and industries of final output. Firms in a common industry with similar final products tend to have a wider spectrum of technological activity than output[8], and each has their own fields of technological specialisation. In most of the relevant empirical work the distinction between technological sectors and industries is fudged. Schmookler, for example, grouped technological categories in accordance with the industry in which their inventions were most likely to be used, as a means of devising an approximate correspondence between the two, and this has become the standard methodology.

When suitably reorganised, patent statistics are rich enough to throw new light on all these three areas of discussion. Pairwise international comparisons can be made by using information on the country of residence of inventors patenting in a third country, which imposes a common legal system and screening procedures[9]. The best choice of third country for this purpose is the USA, as it has the largest market and the most patents granted to foreigners. Sectors of technological activity can be readily developed by using the classifications devised by the patent examiners. Patent classes and sub classes can be grouped by common types of technological activity, rather than by the industries in which they are most likely to be used. For example, there are numerous mechanical technologies which are used in every industry. The industries which actually develop and use technologies can be determined by means of company level data. At the turn of the century just under 20% of US patents were assigned to companies, a share which rose sharply during the interwar

[7] Cf. D. J. Teece/G. Pisano/A. Shuen, Firm capabilities, resources, and the concept of strategy, mimeo, University of California at Berkeley, September 1990. J. Cantwell, The theory of technological competence and its application to international production, in: D. McFetridge (ed.), Foreign Investment, Technology and Economic Growth, Toronto 1992.

[8] Cf. K. Pavitt/M. Robson/J. Townsend, Technological accumulation, diversification and organisation in UK companies, 1945–1983, Science Policy Research Unit Discussion Paper, No. 50, July 1987.

[9] Cf. L. L. G. Soete/S. M. E. Wyatt, The use of foreign patenting as an internationally comparable science and technology output indicator, in: Scientometrics 5, 1983. K. Pavitt, Uses and abuses of patent statistics, University of Sussex Science Policy Research Unit DRC Occasional Paper, No. 41, February 1987.

period, and in recent years has stood at around 80%. In turn, most of these corporate patents are attributable to large firms, whose industries are well known.

The remainder of this paper uses US patent data collected with reference to the country of origin of invention and the company responsible to describe trends in the evolution of European industrial technology in the interwar period. Of course, it does not cover all the issues raised above, which lie beyond the scope of any single paper. However, it does refer to each of the structural aspects of the pattern of technological development that have been outlined, and examines their significance in the context of Europe between the wars. Section 2 provides some further details on the construction of the database. Section 3 begins the investigation of the structural characteristics of technological development in European industries in the 1920s and the 1930s, by comparing the fields of the fastest growth of new technology across countries. Section 4 looks at the pattern of comparative advantage of European countries in technological activity, and at how this changed in the interwar period. Section 5 compares the composition of innovating industries in the largest European countries, by extending the analysis to consider data on firms patenting in the USA. A short conclusion is drawn in section 6.

2. *The Data*

Two types of information have been collected manually from the US Index of Patents and the US Patent Office Gazette, which are available at the UK Patent Office library in London. Firstly, in work currently still in progress, for every patent granted in years between 1890 and 1962 the country of residence of inventor is being recorded. From 1963 this information has already been computerised by the US Patent Office. So far, about two-thirds of the years 1890–1962 have been completed. The evidence on the country of invention discussed below includes the years 1920–24, 1927–29 and 1933–39. As this information on the country of origin of invention relates to individual numbered patents, it can be combined with a sectoral classification of the technological activity with which the patent is associated. This employs the system of patent classes used by the Patent Office; fortunately, as these classes change the Office reclassifies all earlier patents accordingly, so the classification is historically consistent. Although patents may be assigned to several fields, the primary classification was used in all cases. A set of 56 types of technological activity was derived by allocating classes or sub-classes to common groups of activity (see Table 3), and from these 18 broader fields can be identified.

Secondly, using the same sources, all patents assigned to a selection of large firms between 1890 and 1968 were recorded. From 1969 onwards equivalent information has been computerised by the US Patent Office. The firms selected for the historical patent search were identified in one of three ways. The first group consisted of those firms which have accounted for the highest levels of US patenting after 1969; the second group comprised other British, German or US firms which were historically among the largest 200 industrial corporations in each of these countries[10]; and the third group was made up of other companies which featured prominently in the US patent records of earlier years (a method that proved most significant for a number of French firms that had not been identified from other sources). In each case, patents were counted as belonging to a common corporate group where they were assigned to affiliates of a parent company. Affiliate names were normally taken from individual company histories. As lists of individually numbered patents were recorded for each company in every year, it proved possible to match these with the country of origin of invention and the sector of technological activity, in the way just described. So far, lists have been completed for around 150 European corporate groups which were assigned US patents before the Second World War.

3. The Geographical and Sectoral Structure of Technological Development in the Interwar Period

The story which is usually told from an analysis of US patents granted in the interwar period is that patenting accelerated during the 1920s, before reaching a peak in 1932 and then declining to a trough in 1947[11]. This is reflected in the figures for total annual US patenting reported in Table 1. What they show in rather more detail is a phase of postwar recovery between 1920–25; a pause followed by renewed growth from 1926–32; a fall associated with recession from 1933–37; and then a recovery interrupted by the arrival of war after 1938. The equivalent periodisation of underlying technological activity is about two years earlier than each of

[10] Derived from lists in A. D. Chandler, Scale and Scope: The Dynamics of Industrial Capitalism, Cambridge/Mass. 1990.

[11] Cf., for example, M. Brouwer, Schumpeterian Puzzles: Technological Competition and Economic Evolution, Hempstead 1991.

these dates, owing to the lag between invention, patent application and the grant of a patent.

These are average trends, which due to the source used best reflect the path of technological development in the USA rather than other countries. They do not provide any indication of the extent of variation in the growth of patenting across countries or sectors. Some country evidence is provided by Table 2. The first and last rows demonstrate how the growth of total patenting mirrors the US position, which is scarcely surprising as in these years US residents obtained over 80% of US patents granted. In the US case the growth of patenting in the 1920s was similar to that achieved in the pre-war period, but in the 1930s there was a decline in the numbers of patents granted. As a result, the overall growth of patenting in the interwar period was lower than in the years leading up to the First World War. The growth rates stated in Table 2 were calculated by taking the average annual number of patents granted in each of the periods 1890–96, 1920–24, 1927–29 and 1933–39, and then estimating the average annual growth over the span between the mid-years of the relevant periods (1893, 1922, 1928 and 1936).

European patenting in the USA did not slow down in the interwar period in the same way. The number of US patents granted to the residents of European countries grew by an annual average rate of 4% in the interwar years, having (like the patenting of US residents) grown at an annual average rate of about 2% between the early 1890s and early 1920s. Of course, it cannot be concluded from this that technological development was more rapid in Europe than in the USA in the interwar period. There are two reasons why not. Firstly, the European position reflects quite heavily the very strong growth of German based patenting. While there seems little doubt that technological advance did proceed more rapidly in Germany than in most other European countries at this time, in the 1920s the number of US patents originating from German inventions grew at an extraordinary rate of 17% per year, which is artificially high. The immediate postwar decline in German patenting after 1918 was dramatic, as the defeated country whose firms saw their affiliates in the USA expropriated, and which faced economic collapse in the early 1920s.

Secondly, it might be argued that the rise in the share of foreign patenting in the USA in the interwar period simply reflects a greater internationalisation of patenting procedures. This may be related, in turn, to the growth of the US market especially prior to 1929, and the desire of European companies to participate in international cartels that relied on an exchange of patents[12]. Most of these cartels, though, had as their chief objective the sharing of markets rather than agreements to share patents, except in the chemical industry [13]. Cross-licensing arrangements have always been more common in the chemicals industry, and between 1929 and 1948 the major US and British producers, Du Pont and ICI, had an agreement for technological cooperation[14]. In addition, Liebenau has argued that the German chemical companies were especially prone to use patenting as a tool of business strategy[15]. However, if it is true that foreign patenting increased relative to domestic patenting during this period then the position of Canada stands out rather starkly. Canadian patenting in the USA fell throughout the interwar years, having risen steadily at an above average rate of 3% per annum prior to the 1920s. It therefore seems that technological activity was sustained after 1920 much better in both Europe and the USA than in Canada, if Canada can be used as a benchmark for comparisons. Throughout this time, the largest numbers of US patents granted to foreign residents were attributable to inventors living in Britain, Germany or Canada, though the significance of Germany increased markedly relative to the other two.

Perhaps more revealing still is to return to the profile of patenting over time. German patenting in the USA follows the familiar pattern of faster growth in the 1920s than in the 1930s, though this is exaggerated by the very low starting point in the early 1920s for the reasons already mentioned. The same cannot be said, though, for patents granted to British residents. They declined in the 1920s before rising again in the 1930s, precisely the reverse of the general trend. This is consistent with the view that in the interwar period the relative decline of Britain was concentrated in the 1920s, and partly because she was already in a weak position by 1929 she recovered earlier and more surely from the world depression of 1929–33[16]. Although to a lesser extent, a pattern of technological development more similar to the British revival or renewed growth of the 1930s can also be observed for the Netherlands and Sweden. This confirms that it may be

---

12 Cf. R. Liefmann, Cartels, Concerns and Trusts, London 1932; A. Plummer, International Combines in Modern Industry, London 1938; E. Hexner, The International Steel Cartel, Chapel Hill 1943.

13 Cf. Hexner, The International Steel Cartel.

14 Cf. D. A. Hounshell/J. K. Smith, Science and Corporate Strategy: Du Pont R&D, 1902–1980, Cambridge 1988.

15 Cf. J. Liebenau, Patents and the chemical industry: tools of business strategy, in: ibid. (ed.), The Challenge of New Technology: Innovation in British Business since 1850, Aldershot 1988.

16 Cf. D. H. Aldcroft/H. W. Richardson, The British Economy, 1890–1939, London 1969.

dangerous to assume that rates of technological change move together across countries, as is occasionally supposed in the long wave literature.

The sectoral distribution of the growth of technological activity as measured by total US patenting is depicted in Table 3. The broad pattern is predictable, and consistent with other historical evidence on the composition of technological change in the interwar period. That is, the fields of activity which were both significant in terms of their shares of total patenting and were the fastest growing lie in chemical and allied, and electrical technologies, while mechanical technologies were in decline. The significant fields in which patenting grew at over 4% per annum in the interwar period as a whole (compared to the average 0 5%) were on the chemicals and allied side, food and tobacco products, chemical processes, photographic chemistry, cleaning agents and compositions, synthetic resins and fibres, bleaching and dyeing processes, other organic compounds, pharmaceuticals, and coal and petroleum products; and on the electrical side, special radio systems, other electrical communications, and electrical devices and systems.

The fastest growing areas within some of the more general categories can also be established. In food the fastest growing class was sugar, starch and carbohydrates. Within chemical processes, abrading, refrigeration, liquid purification or separation, and coating processes took the lead. Mineral oil processes and products, and hydrocarbons represented the fastest growing areas of coal and petroleum technology. The most rapidly developing electrical devices and systems were conductors and insulators, electron space discharging devices, amplifiers, oscillators, wave transmission, thermal switches, power conversion systems, connectors, and other systems and devices. Within the related area of general electrical equipment, technologies devised for the purposes of refrigeration, and temperature and humidity regulation also increased rapidly.

In some sectors strong technological advance in the interwar period was sustained after 1945, in others it was not. A continued high growth in US patenting after the Second World War is observed in chemical processes, photographic chemistry, synthetic resins and fibres, organic chemicals, pharmaceuticals, and radio systems. However, patents of electrical devices and systems grew at only just over the average rate, while in food and tobacco products, cleaning products and compositions, bleaching and dyeing processes, and coal and petroleum products the shares of US patenting slipped back. In the case of electrical technologies this is because the major interest lay elsewhere, in computing and semiconductors.

There is one other interesting feature of Table 3. The overall growth in patenting in the 1920s followed by a decline in the 1930s as described earlier is not witnessed uniformly across sectors any more than across countries. It is observed in the declining mechanical technologies, and in the electrical rather than chemical fields. In food products, chemical processes, photographic chemistry, and cleaning agents and compositions the growth of technological activity actually increased in the 1930s, while in organic compounds, pharmaceuticals, and coal and petroleum products there was little difference between patent growth rates in the 1920s and 1930s. It is therefore far from clear that the strongest areas of technological development suffered very badly in the recession and its aftermath, even though there was a decline in most other fields of activity, at least in the USA.

Indeed, when examining the US evidence separately as in Table 4 a similar impression emerges. The reversal of the 1930s does not apply to the area of strongest technological development in the interwar period, in chemical activities. By comparison with the overall picture of Table 3, US based patenting in photographic chemicals fell in the 1920s but rose even faster than for other countries in the 1930s; while US originated patents of organic compounds and pharmaceuticals grew more rapidly in the 1930s than in the 1920s, unlike in other countries.

For European based patenting as shown in Table 5 there was a wider range of high growth sectors, as befits the faster overall growth rate which may be at least partly explained by a greater take up of foreign patents. In Europe almost all the chemical fields displayed rapid growth in the interwar period, including inorganic chemicals and agricultural chemicals; and in the electrical area a fast rate of development extended to telecommunications, illumination devices, and other general electrical equipment. However, in Europe there was a greater tendency than in the USA for the rate of increase of chemical technologies to fall in the 1930s by comparison with the 1920s. This is true of food products, cleaning agents and compositions, and pharmaceuticals; and also of photographic chemistry and other organic compounds, although in these fields the growth of European based patenting remained very impressive in the 1930s.

One hypothesis to explain these contrasts might be that while there was a general decline in development in the 1930s, the Europeans had taken the lead in the area of many chemical technologies. Consequently, the US chemicals industry continued with a catching up process in the 1930s despite the decline in activity elsewhere. However, chemical technologies continued to grow rapidly in Europe in the 1930s as well. Related to the European strength in chemicals, there was also a rapid technological development in Europe in chemical machinery and equipment, and in the materials technologies used to create non-metallic mineral products (in the 1920s), and in rubber and plastic products (in the 1930s).

Comparing the largest European countries, Germany emerges as the centre for the most rapid technological development in the interwar period. As measured by US patenting, it witnessed a high rate of technological development in all the chemical and electrical fields, with the sole exception of disinfecting and preserving (a very small category). The details are set out in Table 6. German advances in photographic chemistry, synthetic resins and fibres, and special radio systems were especially spectacular. No doubt related to the development of photographic chemicals, and the links between the new electrical technologies and scientific instruments, the growth of German based patenting in photographic equipment was also high by European standards. Even allowing for the very low base of the early 1920s, it does seem that the rate of technological change in Germany slowed down in the 1930s, in common with most (but not all) other countries. However, there are still some exceptions. In food and tobacco products, for example, German based patenting rose faster in the 1930s than it had done in the 1920s, the reverse of the general European position.

The chemical and electrical fields also constitute the main areas for the growth of British technology in the interwar period as identified in Table 7; the areas in question were food and tobacco products, photographic chemistry, synthetic resins and fibres, bleaching and dyeing, organic compounds, pharmaceuticals, telecommunications, special radio systems, illumination devices, and electrical devices and systems. It is noticeable that this list does not include the coal and petroleum product field, despite the British strength in this area before 1914, which was restored after 1945. However, the major advances in mineral oil refining in the interwar period took place mainly in Germany and the USA, after the early invention of a catalytic process in France in the 1920s[17].

Sectoral disaggregation also helps in exploring the elements underlying the faster growth in technological activity in the UK in the 1930s than in the 1920s, the reverse of the sequence experienced in Germany or other countries. In contrast with Germany, technological development in Britain was enhanced in the 1930s in synthetic resins and fibres, bleaching and dyeing, other organic compounds, and special radio systems. The two sectors that played the greatest role in the increase in British based patenting in the 1930s were organic compounds, and textile and clothing ma-

[17] Cf. C. Freeman, The Economics of Industrial Innovation, London 1982.

chinery. Although textile machinery was an area of traditional British strength, technological activity in this sphere had only been growing slowly since the 1890s. It underwent an interwar revival in both Britain and Germany (unlike in the USA), but in Britain this picked up much more during the 1930s. This is consistent with the argument that there was not a failure of British entrepreneurship within the old established industries themselves, but a failure to switch into the newer science-based industries and thereby to become a follower rather than a leader in technological development in other sectors as well[18]. It should be noted, though, that there were certain branches of chemical technology in which activity fell in Britain in the 1930s while rising in the USA, where chemicals were largely excluded from the general downward trend, as mentioned above. The British decline occured in food and tobacco products, photographic chemistry, and pharmaceuticals, all fields in which American growth was higher in the 1930s.

By comparison with Britain and Germany, the interwar growth in technological activity in France was rather more in electrical than in chemical fields. This can be seen by looking at the figures contained in Table 8. In all but one of the most significant areas of increases in French based patenting, growth was slower in the 1930s than in the 1920s. This holds true for agricultural chemicals, synthetic resins and fibres, pharmaceuticals, coal and petroleum products, telecommunications, special radio systems, and illumination devices. However, in electrical devices and systems, technological activity in France rose in the 1930s, unlike in other countries. The contribution to technological development in France provided by work in telecommunications and illumination devices is shared with Britain and Germany, so it seems that these were fields in which innovation was much more concentrated in Europe than the USA.

4. *The Pattern of Comparative Advantage in Technological Activity in Europe in the Interwar Period*

Following a now familiar procedure[19] an index of national patterns of comparative advantage in technological activity, or revealed technological

[18] Cf. D. N. McCloskey, Economic Maturity and Entrepreneunal Decline: British Iron and Steel, 1870–1913, Cambridge/Mass. 1973; L. G. Sandberg, Lancashire in Decline: A Study in Entrepreneurship, Technology and International Trade, Columbus 1974.

[19] Cf. L. L. G. Soete, The impact of technological innovation on international trade patterns: the evidence reconsidered, in: Research Policy, 16, 1987; J. Cantwell, Technological Innovation and Multinational Corporations, Oxford 1989.

advantage (RTA), can be constructed from patent statistics. This index is measured by the national share of patenting in each sector relative to the national share of total patenting. Let Pij denote the number of US patents in sector i granted for inventions attributable to residents of country j, where the sector i refers to the classification of the technological activity with which the patent is most associated. The value of the RTA index for country j in activity i is then defined as follows:

$$RTA_{ij} = (P_{ij} / \eth_j P_{ij}) / (\eth_i P_{ij} / \eth_i \eth_j P_{ij})$$

For each country the index approximately varies around one, as it represents the country's share of patenting in any given activity relative to its overall share. The index indicates that a country is comparatively advantaged in those sectors in which it has a value greater than unity. In this exercise the index was calculated in two different ways. The first method was used when estimating the RTA of the USA and Europe as a whole, while the other way was employed for individual European countries. In the first case, the position of the USA and Europe respectively were assessed relative to total US patenting. In the second procedure particular European countries were compared with total US patenting traced to all foreign (non-US) locations, excluding patents originating from the inventions of US residents.

One reason for this discrepancy is that it is useful (as was done in the previous section) to begin by comparing Europe as a whole with the USA, before moving on to consider individual European countries by comparison with one another. This is effectively what is entailed by the two varieties of the index, as in the period in question the USA dominates total patenting, while the European countries account for the bulk of what remains. There is, though, a further reason. That is, the determinants of foreign resident patenting in the USA may be somewhat different from the determinants of domestic patenting. Foreign based patents might be expected to include a higher proportion of the most economically significant, and especially historically are likely to vary across sectors according to the extent of foreign penetration of the US market. The use of total foreign patenting in the RTA index for individual European countries therefore comes closer to comparing like with like.

Tables 9 and 10 provide the evidence on the first set of RTA indicators for the USA and Europe. For this purpose the index is shown for a selection of sectors which are typically more broadly defined than in the detailed classification used in earlier tables. The comparative advantage in technological activity of the USA in the interwar period and earlier lay in metal products, mechanical engineering, certain fields of transport equipment and textiles and wood products. The historical roots of American leadership in woodworking technology are well known[20]. In the transport area a more refined disaggregation indicates that US residents were comparatively advantaged in the development of railway equipment, and in vehicles but not engine technologies. In the mechanical fields the US technological advantage was most pronounced in agricultural equipment, construction and excavating equipment, and mining equipment.

By contrast, as today European technological advantage lay principally in the chemical fields. In the chemical areas the USA was comparatively disadvantaged. Although it seems that the relative gap between European and American competence in chemicals was narrower in the interwar period than it had been in the 1890s, it remained wide. The greatest European strengths were centred in agricultural chemicals, bleaching and dyeing and organic compounds. The Europeans also had an advantage in the development of electrical equipment which held up well throughout the interwar period, although in this area the USA had a balanced position (being neither comparatively advantaged nor disadvantaged). Perhaps allied to this, Europe also performed well in the field of professional and scientific instruments, especially in photographic equipment. In the transport equipment area, Europeans were relatively most innovative in motor vehicles, aircraft and shipping technologies. However, in motor vehicles their comparative advantage was in the field of internal combustion engines and not vehicles as such.

There are only two broad areas in which there was a switch between Europe and the USA in comparative advantage in technological activity during the interwar period. One is the field of food and tobacco products, and the other coal and petroleum products. In the latter case at least, the US improvement can be attributed to the emergence of major new developments in mineral oil refining in which US companies played a leading role, as mentioned earlier. Otherwise, the sectoral pattern of technological activity in Europe and the USA remained reasonably stable over the interwar years, as might have been expected from the cumulative nature of technological change[21].

---

20 Cf. N. Rosenberg, Perspectives on Technology, Cambridge 1976.

21 Cf. Cantwell, Historical trends in international patterns of technological innovation (n. 4).

Turning to specific European countries, it needs to be recalled that in their case the RTA index was calculated relative to foreign based patenting alone, and hence mainly relative to one another. Consequently, the comparative advantage of Germany in the technological development of chemicals, as shown in Table 11, is by comparison with Europe as a whole, which as already noted was advantaged in chemicals compared to the USA. The German strength was most significant in organic chemicals, though in this field her comparative advantage declined before 1914, and continued to do so during the interwar period. This is perhaps not surprising given that German firms began from a position of overwhelming superiority in artificial dyestuffs in the latter part of the nineteenth century, and given the major new developments in organic chemistry between the wars. Germans were also strongly represented in the electrical equipment and instruments sectors, or in other words in all the fields of the most significant technological opportunities. By comparison with her position in the 1890s, though, Germany slipped back in motor vehicle technology, in both the vehicle and engine areas.

The technological advantage of the UK in the interwar period, as set out in Table 12, was based mainly on mechanical innovations, and in more traditional technologies such as in textiles and shipbuilding. The greatest British strength was in textile and clothing machinery, but other important fields were assembly and material handling equipment, mining equipment, and printing and publishing machinery. From an average position Britain created and improved upon a comparative advantage in the development of building material processing equipment during the interwar period. As already alluded to, however, she temporarily lost her position (partly to Germany) in coal and petroleum products or oil refining. Although relatively weak in chemicals by German standards between the wars, she did have some strengths in synthetic resins and fibres and bleaching and dyeing processes. British residents were also prominent in patenting in the rubber products field. Overall, however, the UK had a technological profile which was the reverse of the German pattern; British innovation was geared to areas in which technological opportunities had been greater in the previous century rather than the interwar period.

As shown in Table 13, the revealed technological advantage of France between 1920 and 1939 was concentrated in the transport equipment, electrical equipment, professional instruments, and other manufacturing and non-industrial sectors. In the area of electrical technologies, French expertise was especially noteworthy in telecommunications and other electrical communications. In one respect, France occupied a position mid-way between Britain and Germany in terms of the composition of her technological activity. She was relatively well represented in the fast growing areas of electrical communications, instruments and aircraft technology; but her position in chemicals slipped during the interwar period. Indeed, the RTA distribution for France displays a greater mobility or restructuring of activity than for either the UK or Germany, though this is partly due to the smaller scale of French based patenting, perhaps related to the lesser number of large French industrial companies.

Some evidence on the comparative advantage of the smaller but technologically active European countries is given in Tables 14 and 15. As might have been expected, the main Swiss strength lay in chemicals and pharmaceuticals. Switzerland also performed relatively well in electrical and instruments technologies, but it seems that she was subject to something of a falling off in these fields in the 1930s. During the interwar period Swedish technological development favoured the areas of metal products, mechanical engineering, motor vehicles and power plants. In both Switzerland and Sweden, technological evolution between the wars exploited areas of expertise that had emerged before 1914, but a rather more rapid transition into related fields of technological opportunity seems to have been accomplished in Sweden after 1920.

5. The Composition of Innovating Industries in Europe in the Interwar Period

The share of US resident patents assigned to companies rose from under 19% at the turn of the century to over 60% in 1940[22]. The corporate share rose especially fast during the interwar period, and this is associated with the steadily increasing significance of company research laboratories in the USA, as documented by Mowery and Rosenberg[23]. Indeed, the rising share of corporate patenting probably understates the growing significance of corporate research, since large firms have a more cautious patenting strategy and conduct more tests on the viability and usefulness of inventions before patenting [24]. For European resident patents the corporate share has not been fully recorded in the same way, so a direct compar-

[22] Cf. Schmookler, Invention and Economic Growth.

[23] Cf. D. C. Mowery, C. Industrial research and firm size, survival, and growth in American manufacturing, 1921–46, in: Journal of Economic History 43, 1983, pp. 953–80; D. C. Mowery/ N. Rosenberg, Technology and the Pursuit of Economic Growth, Cambridge 1989.

[24] Cf. Schmookler, Invention and Economic Growth.

ison is not possible. However, the US patenting of the largest European industrial firms and their affiliates has been traced, and as a rough guess towards the end of the interwar period they accounted for at least a half of corporate patents from the same country of origin, although of course this varies across countries. The companies whose patenting has been recorded so far are listed in an Appendix.

The share in US patents of these large firms for the inventions of the residents of Germany, the UK, France and Switzerland is shown in Table 16. The firms concerned patented some inventions originating outside their home countries, but for the purpose of this table they have been excluded. The large firm share of German resident patents is therefore to be understood as the share attributable to the domestic research of large German companies. As with the equivalent US corporate share, the general trend is an upward one and (except for France) the increase is especially clear in the interwar period. However, while the corporate share for Germany and Switzerland appears to be roughly similar to the US position, the share for Britain is lower and for France it is much lower. Indeed, even for Germany and Switzerland it seems that the corporate share of patenting increased to a lesser extent in the interwar period than it did for US resident patenting.

Probably the main explanation for this is the slower evolution of the industrial research laboratory in Europe in comparison with the USA. In Britain, in particular, this can be attributed to the weakness of company reorganisation and the ineffectiveness of mergers after the First World War[25]. The restructuring of US industry proved to be a major part of the establishment of in-house research. In France changes in management practice and workplace organisation linked with the establishment of mass production advanced more in a variety of companies than in Britain in the 1920s and 1930s, but production and technology was not as concentrated in the largest firms as it was in Germany, Switzerland or the USA. According to the French economic historian François Caron, in this period French companies relied upon strategies of highly specialised and smaller scale research closely allied to the immediate needs of production.

Differences are also evident in the industrial composition of large firm patenting across the major European countries, which reflects the relative strengths of their companies. The industrial distribution of the US patenting of German firms is set out in Table 17. As explained earlier, in this scheme the patents assigned to a particular company are allocated in their

[25] Cf. Mowery/Rosenberg.

entirety to the primary industry of output of the firm in question. What emerges most clearly is the leading role of the German chemical companies throughout the period, and their absolute dominance of patenting in the 1890s. German electrical companies also developed their position strongly during the interwar period. The concentration of technological activity in Germany in the largest chemical and electrical firms is consistent with the German advantage in the chemical and electrical fields, as shown in Table 11. The shift from chemical to electrical firms between the 1890s and the 1930s probably reflects a change in the relative propensities of the two groups to patent in the USA. However, this change seems to have already occurred by 1920 with the expropriation of a number of the German owned chemical affiliates in the USA; during the interwar period the patent share of the leading German chemical firms increased as rapidly as did that of the electrical firms.

A closer inspection of the company evidence helps to account for certain of the general trends mentioned earlier. One factor behind the reduction in the growth of technological activity in the 1930s in Germany was that the major German firms such as IG Farben were obliged to shift their attentions towards military objectives[26]. Despite this, there is no doubting the great technological strength of the largest German firms during the interwar period. Even Liebenau, who has pointed out that patents are in part a tool of business strategy, has acknowledged that the high and growing share of patents assigned to German chemical firms was a reflection of the relative superiority of the German chemical industry[27]. In any event, the major strategic agreement in the chemicals sector in the 1930s involved Du Pont and ICI, but IG Farben still witnessed the fastest growth of patenting of all, and even the absolute level of its US patenting was on a par with Du Pont. This suggests that even if the German chemical firms were prone to use patents as a tool of business strategy this is in turn because their strong technological base promoted the use of such a strategy. It can be argued that theirs was a "technical capitalism" by contrast with the "commercial capitalism" that persisted in Britain[28].

The pattern of large British firm patenting in the USA underwent much more of a shift than did that of German companies during the interwar period. As can be seen in Table 18, the share of British chemical compa-

---

26 Cf. the contribution of G. Plumpe in this book.

27 Cf. Liebenau, Patents and the chemical industry (n. 15).

28 Cf. the contribution of B. Dornseifer in this book.

nies rose dramatically in the 1930s after the formation of ICI in 1926. This is compatible with the Mowery/Rosenberg hypothesis that the weakness of in-house corporate research in the UK had to do with the absence of effective mergers, and the only mergers of note during the period were those connected with the formation of ICI and Unilever (also principally in chemicals at that time)[29]. It is also related to the Patents and Processes Agreement established in 1929 between ICI and Du Pont[30]. It is certainly the case that prior to 1926 the companies that later joined to form ICI had substantially less corporate research and less US patenting than did ICI in even its early years. The reduction in the rate of growth of research in the 1930s was also less for ICI than for its major international competitors[31], which fits with the improved technological performance of Britain in the 1930s by comparison with the 1920s as described above.

By comparison with other countries, it can also be argued that industrial research in the UK was conducted on a more cooperative basis[32]. However, few US patents were granted to the Research Associations themselves, and it seems that the value of research spending supported by the Department of Scientific and Industrial Research or the Research Associations was small relative to private corporate research, especially by the late 1930s[33]. Yet this should not be taken to imply, as argued by Edgerton and Horrocks, that industrial research was more concentrated in Britain than it was in the USA. The evidence here suggests that corporate research was less concentrated in Britain than the USA (as contended by Mowery/Rosenberg). Industrial R&D in Britain was also less concentrated than in Germany or Switzerland, but more concentrated than in France. This is also consistent with the view of Chandler that Britain lagged the USA and Germany in the adoption of corporate research and formal hierarchy[34]. If the large companies listed in the Appendix accounted for half the US patents assigned to firms from the country in question then the corporate share of patenting for Germany would have been 60% (like in the USA), but for Britain it would have been 45%.

---

29 Cf. Mowery/Rosenberg.

30 Cf. Hounshell/Smith.

31 Cf. D. Edgerton/S. Horrocks, British industrial research and development before 1945, in: Economic History Review, forthcoming 1994.

32 Cf. P. S. Johnson, Cooperative Research in Industry: An Economic Study, London 1973.

33 Cf. Edgerton/Horrocks.

34 Cf. Chandler, Scale and Scope.

Apart from the new role of the major British chemical companies, technological activity was more widely dispersed across industries than in Germany. One area which specifically features in the UK is the rubber products industry. Consistent with the comparative advantage of Britain in mechanical technologies (Table 12), the British mechanical engineering firms performed strongly, but fell back severely in the 1930s. This also accords with the sharp decline of mechanical technologies in the 1930s by comparison with the continued increase in technological activity in the chemicals field during the 1930s, as discussed above.

In France there was also a shift away from technological activity associated with large non-electrical machinery producers, as is revealed by an inspection of Table 19. However, in this case electrical equipment became the most important industry for innovative advance in larger French firms. The rapidly increasing role of the electrical companies perhaps also reflects a rise in the relative propensity of European firms in this industry to patent in the USA by comparison with the much earlier presence of the leading chemical firms in US patenting, as aready discussed in the German context. French based patenting in the interwar period was also even more genuinely widely dispersed across industries than was British based patenting. The motor vehicle, rubber products and building material companies all had an important part in technological activity in France between the wars.

Turning finally to Table 20, US patenting deriving from Switzerland was the most heavily concentrated in large firms of any European country. Relative to the Swiss economy, the biggest chemical firms played a greater role in local technological development than applied even in Germany or the UK. As in Germany, by 1920 there had already been a shift in US patenting away from the chemical producers and towards the electrical equipment companies. However, during the interwar period the chemical firms tended to reassert their position, in line with the generally faster growth of technological activity in the chemical than in the electrical fields in the 1930s.

6. *Conclusion*

It is important to consider the underlying structural characteristics of trends in technological change. The best known trends in the interwar period are the rise of new chemical and electrical technologies, and the overall decline in development in the 1930s relative to the 1920s. a detailed inspection of these trends reveals a more complex picture. Four elements of the structure of technological activity in Europe in the interwar period stand out in particular.

Firstly, the decline in the rate of growth of technological activity in the 1930s which characterised Germany and France as well as the USA did not extend to Britain. This is largely due to a relative British revival in certain branches of chemical and machinery technologies in the 1930s. It might be, of course, that British companies were simply delayed in exploiting the advances which had earlier been taken up by others, and applying them in their own fields of expertise. This requires further investigation.

Secondly, the slower pace of change in the 1930s does not apply to technological activity in the chemicals sector either. Unlike in the electrical fields, the development of chemical technologies continued unabated in the second half of the interwar period. Although this seems to apply particularly in the US case it was also true in Europe, and especially (as already mentioned) in the UK.

Thirdly, contrasting the comparative advantage in technological activity of Britain and Germany in the interwar years helps to account for the faster rate of overall development achieved by Germany. The German advantage was concentrated in the fields of the greatest technological opportunity, namely in chemical and electrical development. This had also been the case in the twenty years or so prior to the First World War[35], but it seems that the discrepancy in the pattern of British and German technological specialisation had become even more pronounced by the interwar period.

Fourthly, technological activity in the interwar period was more industrially concentrated in Germany and Switzerland than in Britain, and much more concentrated than in France. The share of patenting attributable to large firms was higher in Germany and Switzerland, and within this corporate share companies producing chemical products or electrical equipment played a much more prominent role. This may be partly due to the speedier development of the industrial research laboratory in Germany and Switzerland, associated with a greater degree of industrial concentration and managerial reorganisation. However, once again given that the greatest technological opportunities lay in the chemical and electrical fields, this may have contributed to a wider German and Swiss success, especially by comparison with Britain

---

35 Cf. Cantwell, Historical trends in international patterns of technological innovation.

Table 1: The total number of US patents granted annually, 1918–40

Year	Number
1918	38,451
1919	36,796
1920	37,059
1921	37,791
1922	38,360
1923	38,611
1924	42,572
1925	46,432
1926	44,732
1927	41,717
1928	42,357
1929	45,266
1930	45,226
1931	51,755
1932	53,457
1933	48,774
1934	44,419
1935	40,618
1936	39,782
1937	37,683
1938	38,061
1939	43,071
1940	42,237

Source: US patent database developed at the University of Reading with assistance from the US Department of Commerce, Patent and Trademark Office

Table 2: The annual average percentage growth in US patents granted to the residents of various countries over selected periods, 1890–1939

Country	1890–96 to 1920–24	1920–24 to 1927–29	1927–29 to 1933–39	1920–24 to 1933–39
USA	1.90	1.57	−0.74	0.24
Germany	0.47	17.09	4.94	9.99
UK	1.97	−2.77	2.01	−0.07
France	2.76	2.31	1.13	1.63
Italy	6.32	3.18	−1.26	0.62
Netherlands	6.68	7.07	11.17	9.40
Belgium	3.44	2.00	−0.25	0.71
Switzerland	4.55	5.08	1.72	3.14
Sweden	5.37	1.63	3.46	2.67
Austria	−1.80	16.02	2.43	8.05
Canada	3.13	−3.88	−4.49	−4.23
Europe	2.04	5.02	3.24	4.00
total	1.95	1.74	−0.39	0.51

Source: As for Table 1

Table 3: The annual average percentage growth in total US patents over selected periods, classified by fields of technological activity

Sector	1890–96 to 1920–24	1920–24 to 1927–29	1927–29 to 1933–39	1920–24 to 1933–39
1. Food and tobacco products	2.34	3.54	5.24	4.51
2. Distillation processes	6.17	–0.24	6.81	3.73
3. Inorganic chemicals	5.20	–1.39	6.21	2.88
4. Agricultural chemicals	3.04	–4.55	9.74	3.37
5. Chemical processes	4.16	3.99	5.77	5.00
6. Photographic chemistry	4.69	4.32	13.11	9.26
7. Cleaning agents and compositions	3.11	1.62	7.32	4.84
8. Disinfecting and preserving	7.22	–1.27	5.07	2.30
9. Synthetic resins and fibres	6.74	17.40	11.02	13.71
10. Bleaching and dying	2.64	13.08	5.70	8.80
11. Other organic compounds	2.80	13.60	13.43	13.50
12. Pharmaceuticals	2.88	7.93	6.78	7.27
13. Metallurgical processes	4.27	2.91	2.52	2.69
14. Miscellaneous metal products	1.68	0.91	–2.72	–1.18
15. Food and allied equipment	2.49	0.34	1.05	0.75
16. Chemical and allied equipment	1.87	2.39	0.05	1.04
17. Metal working equipment	1.53	–0.72	–2.05	–1.48
18. Paper making apparatus	2.72	5.69	1.99	3.56
19. Building material processing equipment	2.52	2.46	–3.93	–1.24
20. Assembly and material handling equipment	2.25	–0.50	–1.98	–1.35
21. Agricultural equipment	–1.50	–0.64	–5.23	–3.29
22. Construction and excavating equipment	1.69	–1.07	–3.02	–2.19
23. Mining equipment	3.63	6.91	1.04	3.51
24. Electrical lamp manufacturing	2.52	24.56	2.32	11.32
25. Textile and clothing machinery	0.88	3.17	0.53	1.65
26. Printing and publishing machinery	1.19	1.06	0.49	0.73
27. Woodworking tools and machinery	–1.28	5.76	–11.54	–4.51
28. Other specialised machinery	1.92	1.41	–2.36	–0.76
29. Other general industrial equipment	1.44	2.67	–1.50	0.27
30. Mechanical calculators and typewriters	1.90	–4.75	0.51	–1.78

Table 3: (continued)

31. Power plants	1.94	3.02	4.75	4.01
32. Nuclear reactors	0.00	0.00	0.00	0.00
33. Telecommunications	3.73	7.47	0.61	3.49
34. Other electrical communications	–0.07	6.48	4.15	5.14
35. Special radio systems	0.00	24.70	8.40	15.11
36. Image and sound equipment	0.54	10.69	–4.99	1.44
37. Illumination devices	3.82	3.50	1.79	2.52
38. Electrical devices and systems	3.48	11.47	0.04	4.79
39. Other general electrical equipment	3.01	2.67	1.31	1.89
40. Semiconductors	0.00	29.84	0.19	11.97
41. Office equipment	9.28	–6.77	5.16	–0.13
42. Internal combustion engines	8.33	1.74	–4.13	–1.65
43. Motor vehicles	5.75	–4.24	–3.78	–3.97
44. Aircraft	12.61	1.15	–2.54	–0.98
45. Ships and marine propulsion	1.93	–0.64	–7.45	–4.59
46. Railways and railway equipment	–0.10	0.96	–8.40	–4.50
47. Other transport equipment	1.24	–6.66	–3.87	–5.08
48. Textiles, clothing and leather	2.73	1.14	–1.54	–0.40
49. Rubber and plastic products	2.52	–10.30	–0.67	–4.91
50. Non-metallic mineral products	2.44	5.67	1.55	3.29
51. Coal and petroleum products	5.82	13.38	10.94	11.98
52. Photographic equipment	2.11	–1.72	5.64	2.42
53. Other instruments and controls	2.48	0.99	0.23	0.56
54. Wood products	1.07	–1.04	1.25	–2.45
55. Explosive compositions and charges	4.23	–7.45	5.44	–0.29
56. Other manufacturing and non-industrial	1.83	–0.38	–2.00	–1.31
Total	1.95	1.74	–0.39	0.51

Source: As for Table 1

Table 4: The annual average percentage growth in US patents granted to US residents over selected periods, classified by fields of technological activity

Sector	1890–96 to 1920–24	1920–24 to 1927–29	1927–29 to 1933–39	1920–24 to 1933–39
1. Food and tobacco products	2.44	3.23	5.27	4.39
2. Distillation processes	6.43	2.91	7.18	5.33
3. Inorganic chemicals	6.25	−3.72	7.45	2.51
4. Agricultural chemicals	2.47	−5.15	8.64	2.50
5. Chemical processes	4.34	3.87	5.51	4.80
6. Photographic chemistry	4.91	−0.45	14.02	7.58
7. Cleaning agents and compositions	3.17	0.91	7.85	4.82
8. Disinfecting and preserving	7.40	4.13	5.91	5.14
9. Synthetic resins and fibres	7.26	16.50	9.51	12.45
10. Bleaching and dying	1.97	15.03	6.25	9.92
11. Other organic compounds	7.22	8.68	16.51	13.09
12. Pharmaceuticals	3.04	5.44	8.11	6.96
13. Metallurgical processes	4.28	3.10	2.24	2.61
14. Miscellaneous metal products	1.62	1.27	−2.82	−1.09
15. Food and allied equipment	2.01	4.07	0.86	2.23
16. Chemical and allied equipment	2.05	1.53	−0.06	0.62
17. Metal working equipment	1.53	−1.19	−2.27	−1.81
18. Paper making apparatus	2.82	5.30	1.82	3.30
19. Building material processing equipment	2.38	3.29	−4.23	−1.08
20. Assembly and material handling equipment	2.17	−0.73	−2.12	−1.52
21. Agricultural equipment	−1.85	0.04	−5.05	−2.90
22. Construction and excavating equipment	1.50	−0.87	−2.74	−1.94
23. Mining equipment	3.61	7.75	1.03	3.86
24. Electrical lamp manufacturing	2.31	28.22	−0.45	10.95
25. Textile and clothing machinery	0.92	2.65	0.23	1.26
26. Printing and publishing machinery	1.19	0.05	0.52	0.32
27. Woodworking tools and machinery	−1.85	7.99	−12.33	−4.14
28. Other specialised machinery	1.87	1.61	−2.67	−0.86
29. Other general industrial equipment	1.32	3.06	−1.61	0.37
30. Mechanical calculators and typewriters	1.55	−3.70	−0.15	−1.69

Table 4: (continued)

31. Power plants	1.76	1.61	4.90	3.48
32. Nuclear reactors	0.00	0.00	0.00	0.00
33. Telecommunications	3.85	5.56	0.08	2.39
34. Other electrical communications	–0.39	7.49	3.66	5.28
35. Special radio systems	0.00	25.10	6.25	13.95
36. Image and sound equipment	0.40	12.14	–5.97	1.40
37. Illumination devices	4.01	2.81	–0.01	1.19
38. Electrical devices and systems	3.35	11.36	–0.60	4.36
39. Other general electrical equipment	3.10	1.73	1.65	1.69
40. Semiconductors	0.00	32.91	–5.10	9.64
41. Office equipment	9.15	–7.93	4.36	–1.10
42. Internal combustion engines	9.03	2.48	–5.31	–2.04
43. Motor vehicles	5.51	–4.26	–3.99	–4.11
44. Aircraft	12.32	1.97	–3.28	–1.06
45. Ships and marine propulsion	1.81	0.47	–8.00	–4.46
46. Railways and railway equipment	–0.16	–0.61	–8.14	–4.49
47. Other transport equipment	1.12	–6.72	–4.05	–5.20
48. Textiles, clothing and leather	2.68	1.59	–1.75	–0.33
49. Rubber and plastic products	6.25	–11.25	–1.87	–6.01
50. Non-metallic mineral products	2.56	5.12	1.60	3.10
51. Coal and petroleum products	5.96	14.36	10.94	12.40
52. Photographic equipment	2.18	–1.34	0.59	–0.23
53. Other instruments and controls	2.52	0.20	0.37	0.30
54. Wood products	0.95	–3.46	–1.34	–2.25
55. Explosive compositions and charges	4.32	–3.37	3.92	0.73
56. Other manufacturing and non-industrial	1.83	–0.40	–2.03	–1.33
Total	1.90	1.57	–0.74	0.24

Source: As for Table 1

Table 5: The annual average percentage growth in US patents granted to European residents over selected periods, classified by fields of technological activity

Sector	1890–96 to 1920–24	1920–24 to 1927–29	1927–29 to 1933–39	1920–24 to 1933–39
1. Food and tobacco products	0.75	8.70	5.24	6.70
2. Distillation processes	5.07	–3.36	5.54	1.63
3. Inorganic chemicals	2.27	7.09	3.12	4.80
4. Agricultural chemicals	4.56	–0.65	12.06	6.42
5. Chemical processes	2.89	5.81	7.64	6.85
6. Photographic chemistry	3.21	22.22	11.91	16.22
7. Cleaning agents and compositions	2.10	9.11	4.78	6.62
8. Disinfecting and preserving	4.41	–16.73	6.97	–3.92
9. Synthetic resins and fibres	5.07	21.65	18.68	19.95
10. Bleaching and dying	3.54	11.38	5.45	7.95
11. Other organic compounds	–0.24	21.26	10.06	14.73
12. Pharmaceuticals	2.07	19.15	2.76	9.49
13. Metallurgical processes	3.73	3.38	4.95	4.27
14. Miscellaneous metal products	1.80	–1.23	0.08	–0.48
15. Food and allied equipment	3.24	–11.01	3.26	–3.11
16. Chemical and allied equipment	–0.19	12.34	1.61	6.07
17. Metal working equipment	1.41	5.46	1.44	3.14
18. Paper making apparatus	1.44	9.29	4.82	6.71
19. Building material processing equipment	3.19	–3.45	–0.10	–1.55
20. Assembly and material handling equipment	3.73	4.00	1.19	2.38
21. Agricultural equipment	4.20	–3.44	–5.65	–4.71
22. Construction and excavating equipment	5.53	–5.35	–1.63	–3.24
23. Mining equipment	4.41	–10.91	3.89	–2.73
24. Electrical lamp manufacturing	2.59	14.24	11.93	12.91
25. Textile and clothing machinery	0.51	6.93	2.52	4.39
26. Printing and publishing machinery	1.30	9.44	0.90	4.48
27. Woodworking tools and machinery	2.59	–4.88	–3.32	–3.99
28. Other specialised machinery	2.03	1.50	2.08	1.83
29. Other general industrial equipment	2.17	0.90	0.21	0.50
30. Mechanical calculators and typewriters	4.49	–10.96	5.57	–1.86

Table 5: (continued)

31. Power plants	2.74	8.89	5.07	6.69
32. Nuclear reactors	0.00	0.00	0.00	0.00
33. Telecommunications	3.44	24.39	3.08	11.73
34. Other electrical communications	2.91	0.74	8.76	5.25
35. Special radio systems	0.00	23.33	14.06	17.94
36. Image and sound equipment	0.71	6.61	1.19	3.48
37. Illumination devices	1.13	13.15	13.18	13.17
38. Electrical devices and systems	4.65	13.84	5.18	8.81
39. Other general electrical equipment	1.75	12.06	–0.34	4.80
40. Semiconductors	0.00	16.50	25.47	21.54
41. Office equipment	10.00	3.10	10.92	7.50
42. Internal combustion engines	6.03	–2.10	1.98	0.21
43. Motor vehicles	10.50	–3.19	–0.91	–1.89
44. Aircraft	14.33	–1.15	0.45	–0.24
45. Ships and marine propulsion	2.03	–3.97	–3.53	–3.72
46. Railways and railway equipment	0.09	7.68	–9.74	–2.65
47. Other transport equipment	2.49	–4.52	–1.09	–2.57
48. Textiles, clothing and leather	2.60	–4.69	4.89	0.67
49. Rubber and plastic products	0.65	0.93	6.60	4.13
50. Non-metallic mineral products	0.63	13.20	2.57	7.00
51. Coal and petroleum products	5.46	8.20	11.10	9.84
52. Photographic equipment	0.94	–0.31	20.56	11.13
53. Other instruments and controls	2.19	8.18	–0.06	3.39
54. Wood products	0.65	–7.06	2.98	–1.45
55. Explosive compositions and charges	3.70	–20.96	14.72	–2.21
56. Other manufacturing and non-industrial	1.63	–0.06	–0.81	–0.49
Total	2.04	5.02	3.24	4.00

Source: As for Table 1

Table 6: The annual average percentage growth in US patents granted to German residents over selected periods, classified by fields of technological activity

Sector	1890–96 to 1920–24	1920–24 to 1927–29	1927–29 to 1933–39	1920–24 to 1933–39
1. Food and tobacco products	1.34	3.93	9.64	7.15
2. Distillation processes	4.76	13.35	4.37	8.13
3. Inorganic chemicals	1.37	28.45	3.54	13.56
4. Agricultural chemicals	6.12	3.79	12.79	8.84
5. Chemical processes	2.30	9.78	12.22	11.17
6. Photographic chemistry	−1.75	52.90	15.84	30.47
7. Cleaning agents and compositions	0.94	15.77	9.66	12.23
8. Disinfecting and preserving	5.07	−100.00	n. a.	−2.37
9. Synthetic resins and fibres	6.41	29.03	19.31	23.39
10. Bleaching and dying	4.10	11.08	8.56	9.63
11. Other organic compounds	−2.08	31.18	9.28	18.18
12. Pharmaceuticals	3.85	15.25	4.99	9.27
13. Metallurgical processes	3.56	21.47	5.16	11.86
14. Miscellaneous metal products	−1.03	10.92	0.29	4.72
15. Food and allied equipment	1.29	0.00	3.43	1.94
16. Chemical and allied equipment	−1.48	21.39	3.83	11.02
17. Metal working equipment	−0.04	20.19	1.62	9.20
18. Paper making apparatus	0.78	17.97	5.17	10.48
19. Building material processing equipment	−0.57	15.17	−0.65	5.84
20. Assembly and material handling equipment	2.15	20.01	2.50	9.67
21. Agricultural equipment	1.61	8.89	−9.18	−1.84
22. Construction and excavating equipment	3.96	−1.58	0.00	−0.68
23. Mining equipment	5.07	−16.73	3.19	−5.87
24. Electrical lamp manufacturing	1.17	30.77	17.25	22.86
25. Textile and clothing machinery	0.58	14.87	4.06	8.56
26. Printing and publishing machinery	0.84	25.55	0.97	10.85
27. Woodworking tools and machinery	1.54	−2.99	−0.73	−1.71
28. Other specialised machinery	−0.04	11.39	2.75	6.37
29. Other general industrial equipment	−0.26	15.08	0.96	6.79
30. Mechanical calculators and typewriters	7.37	−9.54	9.29	−0.78

Table 6: (continued)

31. Power plants	0.76	42.39	3.41	18.60
32. Nuclear reactors	0.00	0.00	0.00	0.00
33. Telecommunications	3.73	41.45	2.78	17.86
34. Other electrical communications	2.25	–1.94	19.95	10.02
35. Special radio systems	0.00	46.78	22.28	32.24
36. Image and sound equipment	–2.08	15.52	3.08	8.24
37. Illumination devices	–0.15	30.51	16.78	22.48
38. Electrical devices and systems	2.70	25.82	6.57	14.43
39. Other general electrical equipment	1.66	21.85	1.10	9.52
40. Semiconductors	0.00	n. a.	32.38	n. a.
41. Office equipment	11.32	8.89	13.93	11.74
42. Internal combustion engines	1.09	18.83	3.47	9.79
43. Motor vehicles	3.81	2.21	8.90	5.98
44. Aircraft	16.64	7.99	–5.44	0.10
45. Ships and marine propulsion	2.00	6.57	–6.07	–0.85
46. Railways and railway equipment	–2.50	25.89	–10.58	3.54
47. Other transport equipment	–0.54	3.14	6.71	5.16
48. Textiles, clothing and leather	0.51	3.99	6.03	5.15
49. Rubber and plastic products	–0.08	16.78	9.84	12.76
50. Non-metallic mineral products	–1.04	21.80	5.08	11.95
51. Coal and petroleum products	12.55	22.68	12.48	16.74
52. Photographic equipment	1.17	5.23	26.52	16.91
53. Other instruments and controls	1.30	17.90	–0.03	7.29
54. Wood products	–1.72	–3.86	4.12	0.62
55. Explosive compositions and charges	3.61	–28.04	25.10	–1.29
56. Other manufacturing and non-industrial	1.10	9.00	–2.42	2.32
Total	0.47	17.09	4.94	9.99

Source: As for Table 1

n. a. = Not applicable

Table 7: The annual average percentage growth in US patents granted to British residents over selected periods, classified by fields of technological activity

Sector	1890–96 to 1920–24	1920–24 to 1927–29	1927–29 to 1933–39	1920–24 to 1933–39
1. Food and tobacco products	0.63	8.12	4.79	6.20
2. Distillation processes	5.63	−8.28	−7.98	−8.11
3. Inorganic chemicals	0.58	−6.53	3.01	−1.19
4. Agricultural chemicals	2.18	−13.58	14.72	1.61
5. Chemical processes	1.48	0.24	6.97	4.03
6. Photographic chemistry	5.07	13.01	1.36	6.20
7. Cleaning agents and compositions	1.90	−0.42	2.19	1.06
8. Disinfecting and preserving	0.00	−100.00	n. a.	−2.37
9. Synthetic resins and fibres	3.14	2.95	38.17	21.80
10. Bleaching and dying	3.46	7.25	9.46	8.51
11. Other organic compounds	3.87	7.93	20.14	14.75
12. Pharmaceuticals	−0.60	14.23	3.19	7.79
13. Metallurgical processes	3.99	−12.66	4.90	−3.02
14. Miscellaneous metal products	2.23	−7.98	−1.11	−4.12
15. Food and allied equipment	4.44	−19.85	6.97	−5.48
16. Chemical and allied equipment	−0.80	8.17	0.08	3.47
17. Metal working equipment	0.78	−2.55	0.34	−0.91
18. Paper making apparatus	1.38	3.68	4.80	4.32
19. Building material processing equipment	3.32	−6.27	0.22	−2.61
20. Assembly and material handling equipment	3.73	−0.92	−1.25	−1.11
21. Agricultural equipment	5.52	−13.81	−0.73	−6.57
22. Construction and excavating equipment	5.46	−10.91	−5.37	−7.79
23. Mining equipment	3.05	−10.41	1.14	−3.98
24. Electrical lamp manufacturing	0.00	−100.00	n. a.	4.94
25. Textile and clothing machinery	0.07	2.13	3.82	3.09
26. Printing and publishing machinery	1.97	−2.73	0.37	−0.97
27. Woodworking tools and machinery	1.91	−26.42	14.72	−5.16
28. Other specialised machinery	1.88	−5.09	1.77	−1.23
29. Other general industrial equipment	2.27	−7.91	−0.33	−3.65
30. Mechanical calculators and typewriters	0.49	−13.97	5.81	−3.17

Table 7: (continued)

31. Power plants	4.46	−10.15	10.67	1.22
32. Nuclear reactors	0.00	0.00	0.00	0.00
33. Telecommunications	3.31	20.94	1.08	9.15
34. Other electrical communications	2.59	−11.33	10.00	0.29
35. Special radio systems	0.00	7.17	12.53	10.20
36. Image and sound equipment	3.39	1.93	−1.91	−0.28
37. Illumination devices	0.52	5.16	5.13	5.15
38. Electrical devices and systems	4.66	11.40	0.15	4.82
39. Other general electrical equipment	0.84	−0.84	2.62	1.12
40. Semiconductors	0.00	n. a.	10.00	n. a.
41. Office equipment	8.32	3.99	6.19	5.24
42. Internal combustion engines	6.57	−14.30	−0.03	−6.42
43. Motor vehicles	0.00	−7.78	−5.25	−6.34
44. Aircraft	18.41	−10.30	1.52	−3.72
45. Ships and marine propulsion	1.80	−10.72	−2.43	−6.07
46. Railways and railway equipment	−0.76	3.40	−14.69	−7.36
47. Other transport equipment	2.40	−9.76	−7.41	−8.42
48. Textiles, clothing and leather	2.52	−6.97	1.40	−2.28
49. Rubber and plastic products	−0.04	−10.12	9.82	0.78
50. Non-metallic mineral products	0.13	10.11	−1.29	3.45
51. Coal and petroleum products	5.17	−7.40	3.19	−1.49
52. Photographic equipment	−0.81	−4.52	17.19	7.34
53. Other instruments and controls	1.76	2.28	−0.85	0.48
54. Wood products	0.74	−8.51	−2.91	−5.35
55. Explosive compositions and charges	1.17	−4.88	7.78	2.16
56. Other manufacturing and non-industrial	1.69	−7.38	−0.85	−3.70
Total	1.97	−2.77	2.01	−0.07

Source: As for Table 1

Table 8: The annual average percentage growth in US patents granted to French residents over selected periods, classified by fields of technological activity

Sector	1890–96 to 1920–24	1920–24 to 1927–29	1927–29 to 1933–39	1920–24 to 1933–39
1. Food and tobacco products	–0.34	11.96	–5.68	1.51
2. Distillation processes	1.17	–13.58	23.95	6.20
3. Inorganic chemicals	3.32	–10.22	8.42	0.00
4. Agricultural chemicals	0.00	8.89	6.97	7.79
5. Chemical processes	3.61	1.16	3.51	2.49
6. Photographic chemistry	0.00	22.22	–3.88	6.54
7. Cleaning agents and compositions	3.81	10.72	–2.81	2.77
8. Disinfecting and preserving	0.00	0.00	0.00	0.00
9. Synthetic resins and fibres	2.18	24.64	4.56	12.74
10. Bleaching and dying	–0.54	–4.52	8.25	2.58
11. Other organic compounds	2.85	2.11	3.69	3.01
12. Pharmaceuticals	0.17	35.21	–5.74	10.03
13. Metallurgical processes	1.17	3.32	3.03	3.15
14. Miscellaneous metal products	3.25	–0.57	–0.17	–0.34
15. Food and allied equipment	0.53	–16.73	10.00	–2.38
16. Chemical and allied equipment	1.44	13.58	0.29	5.78
17. Metal working equipment	2.74	6.70	–5.52	–0.47
18. Paper making apparatus	0.39	–5.13	5.98	1.07
19. Building material processing equipment	3.19	–10.30	10.00	0.79
20. Assembly and material handling equipment	4.06	1.27	5.37	3.59
21. Agricultural equipment	2.72	1.16	–7.77	–4.04
22. Construction and excavating equipment	2.59	–9.33	21.39	7.12
23. Mining equipment	7.61	–100.00	n. a.	0.49
24. Electrical lamp manufacturing	–100.00	n. a.	–17.52	n. a.
25. Textile and clothing machinery	–0.58	13.14	–4.90	2.45
26. Printing and publishing machinery	–3.43	22.22	–0.65	8.57
27. Woodworking tools and machinery	5.80	–1.58	–14.50	–9.18
28. Other specialised machinery	3.33	7.05	–2.99	1.19
29. Other general industrial equipment	2.99	1.16	0.38	0.71
30. Mechanical calculators and typewriters	5.07	–16.73	0.00	–7.55

Table 8: (continued)

31. Power plants	−2.59	−9.33	16.65	4.71
32. Nuclear reactors	0.00	0.00	0.00	0.00
33. Telecommunications	0.60	24.51	2.61	11.48
34. Other electrical communications	7.91	2.41	0.00	1.03
35. Special radio systems	0.00	30.77	2.83	13.99
36. Image and sound equipment	1.17	4.17	0.31	1.95
37. Illumination devices	1.78	12.25	3.48	7.15
38. Electrical devices and systems	5.22	−1.14	9.50	4.80
39. Other general electrical equipment	1.23	10.94	−2.97	2.77
40. Semiconductors	0.00	8.89	12.53	10.95
41. Office equipment	0.00	−11.63	16.65	3.56
42. Internal combustion engines	7.95	−2.99	3.00	0.39
43. Motor vehicles	15.18	−8.64	2.90	−2.21
44. Aircraft	14.79	−7.14	11.89	3.30
45. Ships and marine propulsion	1.56	−5.73	0.22	−2.38
46. Railways and railway equipment	2.14	−5.31	3.72	−0.25
47. Other transport equipment	4.95	−8.80	−3.01	−5.53
48. Textiles, clothing and leather	5.24	−14.94	13.44	0.27
49. Rubber and plastic products	1.81	4.91	−3.53	0.00
50. Non-metallic mineral products	1.44	8.89	4.41	6.31
51. Coal and petroleum products	−1.98	44.67	0.64	17.57
52. Photographic equipment	1.35	−10.91	20.20	5.72
53. Other instruments and controls	3.80	3.57	0.45	1.78
54. Wood products	6.12	−5.90	3.19	0.81
55. Explosive compositions and charges	7.61	−28.04	−1.91	−14.10
56. Other manufacturing and non industrial	2.31	−5.38	−0.47	−2.61
Total	2.76	2.31	1.13	1.63

Source: As for Table 1

n. a. = Not applicable.

Table 9: The index of revealed technological advantage for the USA for selected sectors and periods, 1890–1939

Sector	1890–96	1920–24	1927–29	1933–39
Food and tobacco products	0.94	0.98	0.97	1.00
Chemicals nes	0.72	0.88	0.80	0.84
Pharmaceuticals	0.88	0.93	0.82	0.93
Metal products	1.02	1.02	1.05	1.06
Mechanical engineering nes	1.01	1.00	1.01	1.03
Electrical equipment nes	0.98	1.00	0.98	0.96
Motor vehicles	0.94	0.95	0.97	0.94
Aircraft	0.92	0.87	0.92	0.89
Other transport equipment	1.03	1.01	1.01	1.04
Textiles and wood products	1.03	1.02	1.06	1.07
Non-metallic mineral products	0.99	1.04	1.02	1.05
Coal and petroleum products	0.89	0.94	0.99	1.02

Source: As for Table 1

Table 10: The index of revealed technological advantage for Europe for selected sectors and periods, 1890–1939

Sector	1890–96	1920–24	1927–29	1933–39
Food and tobacco products	1.78	1.10	1.22	0.92
Chemicals nes	4.50	2.41	2.98	2.21
Pharmaceuticals	2.33	1.81	2.70	1.49
Metal products	0.72	0.75	0.57	0.57
Mechanical engineering nes	0.90	0.95	0.88	0.81
Electrical equipment nes	1.24	1.04	1.21	1.28
Motor vehicles	1.74	1.61	1.20	1.39
Aircraft	1.73	2.62	1.88	1.80
Other transport equipment	0.64	0.79	0.77	0.65
Textiles and wood products	0.67	0.62	0.39	0.45
Coal and petroleum products	2.09	1.84	1.15	0.87
Professional and scientific instruments	1.19	1.04	1.26	1.07

Source: As for Table 1

Table 11: The index of revealed technological advantage for Germany for selected sectors and periods, 1890–1939

Sector	1890–96	1920–24	1927–29	1933–39
Chemicals nes	1.90	1.86	1.58	1.42
Pharmaceuticals	1.23	3.40	1.42	1.37
Metal products	1.02	0.72	0.85	0.81
Mechanical engineering nes	0.87	0.86	0.86	0.81
Electrical equipment nes	1.14	1.19	1.14	1.14
Motor vehicles	1.36	0.53	0.67	0.74
Other transport equipment	0.79	0.63	0.71	0.68
Coal and petroleum products	0.13	0.61	1.66	1.50
Professional and scientific instruments	0.97	1.32	1.07	1.01

Source: As for Table 1

Table 12: The index of revealed technological advantage for the UK for selected sectors and periods, 1890–1939

Sector	1890–96	1920–24	1927–29	1933–39
Food and tobacco products	0.68	0.56	0.93	0.91
Chemicals nes	0.64	0.74	0.63	0.85
Pharmaceuticals	0.61	0.31	0.37	0.39
Metal products	0.94	1.00	0.95	0.95
Mechanical engineering nes	1.10	1.09	1.12	1.21
Electrical equipment nes	0.91	0.96	1.10	0.92
Motor vehicles	1.00	1.33	0.99	0.86
Aircraft	0.33	1.15	0.96	1.07
Other transport equipment	0.92	0.86	0.93	0.78
Textiles and wood products	1.09	0.86	1.20	1.12
Rubber products	1.68	1.10	0.98	1.43
Coal and petroleum products	1.18	1.32	0.77	0.43

Source: As for Table 1

Table 13: The index of revealed technological advantage for France for selected sectors and periods, 1890–1939

Sector	1890–96	1920–24	1927–29	1933–39
Chemicals nes	1.03	1.16	0.75	0.56
Pharmaceuticals	0.91	0.46	1.12	0.62
Metal products	0.88	0.80	0.99	1.10
Mechanical engineering nes	0.92	0.84	1.04	1.04
Electrical equipment nes	1.33	1.05	1.00	0.94
Motor vehicles	1.03	1.63	1.48	1.99
Aircraft	1.21	1.39	1.05	2.72
Other transport equipment	0.85	1.21	0.86	1.60
Rubber products	1.12	1.00	1.66	0.92
Coal and petroleum products	1.82	0.21	1.32	0.65
Professional and scientific instruments	1.00	1.29	1.06	1.24
Other manufacturing and non-industrial	1.45	1.57	1.17	1.35

Source: As for Table 1

Table 14: The index of revealed technological advantage for Switzerland for selected sectors and periods, 1890–1939

Sector	1890–96	1920–24	1927–29	1933–39
Chemicals nes	2.98	2.18	2.05	1.62
Pharmaceuticals	0.00	1.08	2.05	1.40
Metal products	0.78	0.67	0.77	0.80
Mechanical engineering nes	0.63	0.96	0.79	0.88
Electrical equipment nes	1.33	1.30	1.15	0.86
Professional and scientific instruments	2.47	1.38	1.11	0.81

Source: As for Table 1

Table 15: The index of revealed technological advantage for the Sweden for selected sectors and periods, 1890–1939

Sector	1890–96	1920–24	1927–29	1933–39
Chemicals nes	0.41	0.87	0.67	0.37
Metal products	0.74	0.92	1.15	1.32
Mechanical engineering nes	1.25	1.19	1.20	1.44
Power plants	0.00	1.96	2.96	2.54
Electrical equipment nes	0.75	0.68	1.01	0.82
Motor vehicles	0.00	1.14	1.30	1.37
Aircraft	0.00	0.00	0.63	0.94
Other transport equipment	0.82	1.11	0.92	0.88
Professional and scientific instruments	1.07	1.12	0.80	0.89

Source: As for Table 1

Table 16: The corporate share of US patents granted to the residents of European countries for selected large firms, 1890–1939.(%)

	1890 99	1920–29	1930–39
Germany	11.56	22.05	31.34
UK	0.98	7.08	22.70
France	1.41	8.54	8.94
Switzerland	19.76	20.49	43.68

Source: As for Table 1.

Table 17: The industrial shares of the US patenting of selected large German firms, 1890–1939.(%)

Industry	1890–99	1920–29	1930–39
Chemicals and pharmaceuticals	73.21	42.34	54.06
Metals	6.24	17.57	5.02
Mechanical engineering	2.59	3.60	3.83
Electrical equipment	16.59	20.63	26.87
Motor vehicles	0.00	3.74	4.53
Aircraft	0.00	4.73	0.36
Textiles and Clothing	0.00	0.00	0.03
Professional and scientific instruments	1.37	7.39	5.30
Total	100.00	100.00	100.00

Source: As for Table 1.

Table 18: The industrial shares of the US patenting of selected large British firms, 1890–1939.(%)

Industry	1890–99	1920–29	1930–39
Drink	0.00	0.00	0.59
Chemicals and pharmaceuticals	33.78	9.05	32.04
Metals	0.00	2.77	4.17
Mechanical engineering	51.36	36.31	6.06
Electrical equipment	14.86	15.20	16.91
Office equipment	0.00	0.48	0.14
Motor vehicles	0.00	6.39	3.64
Aircraft	0.00	1.09	0.95
Other transportation equipment	0.00	0.72	0.00
Textiles and clothing	0.00	11.22	23.77
Printing and publishing	0.00	0.00	0.95
Rubber and plastic products	0.00	7.72	6.44
Non-metallic mineral products	0.00	8.08	2.52
Coal and petroleum products	0.00	0.97	0.80
Professional and scientific instruments	0.00	0.00	1.02
Total	100.00	100.00	100.00

Source: As for Table 1.

Table 19: The industrial shares of the US patenting of selected large French firms, 1890–1939.(%)

Industry	1890–99	1920–29	1930–39
Chemicals and pharmaceuticals	34.48	23.33	16.05
Metals	0.00	12.05	9.78
Mechanical engineering	51.72	43.33	10.96
Electrical equipment	0.00	4.36	32.09
Motor vehicles	0.00	8.46	6.65
Aircraft	0.00	2.05	2.15
Rubber and plastic products	13.80	3.08	8.82
Non-metallic mineral products	0.00	3.34	13.50
Total	100.00	100.00	100.00

Source: As for Table 1.

Table 20: The industrial shares of the US patenting of selected large Swiss firms, 1890–1939.(%)

Industry	1890–99	1920–29	1930–39
Chemicals and pharmaceuticals	100.00	54.79	59.45
Mechanical engineering	0.00	9.83	13.12
Electrical equipment	0.00	35.38	27.43
Total	100.00	100.00	100.00

Source: As for Table 1.

Appendix

A list of large companies recorded as patenting in the USA by country and sector

Germany

1. Chemicals and Pharmaceuticals
   Cassella & Co. (*)
   Hoechst (*)
   Kalle & Co. (*)
   Bayer (*)
   AGFA *)
   BASF (*)
   IG Farben
   J. D. Riedel & De Haën Chemische Fabrik
   Consortium für Elektrochemische Industrie
   Schering
   Elektrochemische Werke
   Von Heyden
   C. F. Boehringer & Söchnc
   R. Koepp Chemische Fabrik
   E. Merck
   DEGUSSA
2. Metals
   Krupp
   W. C. Heraeus
   Felten & Guilleaume Carlswerke
   Th. Goldschmidt
   Mannesmannröhrenwerke
   Deutsche Edestahlwerke
   Metallgesellschaft
   Thyssen & Co
   Hartmann & Braun
3. Mechanical Engineering
   Maschinenfabrik Augsburg Nürnberg (MAN)
   Rheinische Metallwaren und Maschinenfabrik

Demag
Gesellschaft für Linde's Eismaschinen
Gutehoffnungshütte Oberhausen
Klöckner Humboldt Deutz
Motorenfabrik Deutz
Friedrich Deckel
Julius Pintsch
Schubert & Salzer Maschinenfabrik

4. Electrical Equipment
Siemens & Halske AG
Bergmann Elektricitäts Werke
AEG
Telefunken

5. Motor Vehicles
Daimler
Mercedes Büromaschinen
Bosch
Zahnradfabrik Friedrichshafen

6. Aircraft
Luftschiffbau Zeppelin Stiftung

7. Textiles and Clothing
Vereignigte Glanzstoff Fabriken

8. Professional and Scientific Instruments
Carl Zeiss
Zeiss Ikon
C. P. Goerz
Ernst Leitz Optische Werke

(*) Joined in the formation of IG Farben in 1925

UK

1. Drink
Distillers
J. Lyons

2. Chemicals and Pharmaceuticals
British Dyestuffs Corporation (*)
Scottish Dyes (*)
Levinstein (*)
United Alkali (*)
Nobel Explosives (*)
British Cyanides (*)
ICI
Lever Brothers
Unilever
BOC

3. Metals
   J. Stone & Co.
   Callender's Cables & Construction
   Telegraph Construction & Maintenance Co.
   Metal Box
   W. T. Henley's Telegraph Works
   English Steel Corp.
   Stewarts & Lloyds
   Borax Consolidated
   Johnson Matthey & Co.

4. Mechanical Engineering
   Vickers
   Thos. Firth & John Brown
   Platt Brothers
   Mather & Platt
   Dobson & Barlow
   Howard & Bullough

5. Electrical Equipment
   Metropolitan Vickers Electric Co.
   Associated Electrical Industries
   English Electric
   Marconi
   D. Napier & Sons
   EMI
   Gramophone Co.
   A. Reyrolle & Co.
   Plessey
   C. A. Parsons & Co.
   M-O Valve Co.
   Siemens Brothers & Co.

6. Office Equipment
   British Tabulating Machine
   Imperial Typewriter Co.

7. Motor Vehicles
   Armstrong-Siddeley Motors
   Rolls Royce
   Bentley Engineering Co.
   Austin Motor Co.
   Guy Motors
   Lucas
   Automotive Products

8. Aircraft
   British Aircraft Corp.
   Bristol Aeroplane Co.

De Havilland Aircraft Co.
Fairey Aviation

9. Other Transportation Equipment
Birmingham Small Arms Co.

10. Textiles and Clothing
Courtaulds
Calico Printers Association
British Celanese and H. & C. Dreyfus
Tootal Broadhurst Lee Co.

11. Paper Products
E. S. & A. Robinson

12. Printing and Publishing
Bell Punch Co.

13. Rubber and Plastic Products
Dunlop

14. Non-Metallic Mineral Products
Pilkington Brothers
Morgan Crucible Co.

15. Coal and Petroleum Products
Anglo-Iranian Oil Co. (BP)
Shell Oil Co.
Asiatic Petroleum Co.

16. Professional and Scientific Instruments
Ilford
Cambridge Instrument Co.

(*) Joined in the formation of ICI in 1926.

France

1. Chemicals and Pharmaceuticals
L'Air Liquide
Ugine
Societé Chimique de la Grande Paroisse
Societé des Matieres Colorantes et Produits Chimiques de Saint Denis
Sté Chimique des Usines du Rhone (Rhone Poulenc)
Compagnie de Produits Chimiques d'Alais, Froges et Camargue
Compagnie pour la Fabrication de Material d'Usines á Gaz
Compagnie Nationale des Matieres Colorantes et Manufactures des Products Chimiques du Nord Reunies Etablissements Kuhlmann

2. Metals
Compagnie de Forges et Acierier de la Marine et Homecourt

3. Mechanical Engineering
   Schneider & Cie
   Societé Alsacienne de Constructions Mecaniques (Alsthom)

4. Electrical Equipment
   Compagnie Generale d'Electricité
   Compagnie Generale de Telegraphie Sans Fils

5. Motor Vehicles
   Societé du Carburateur Zenith
   Societé Anonyme des Aciens Etablissements Panhard & Levassor
   SA des Automobiles Peugeot
   SA André Citroen
   SA des Usines Chausson
   SA Française du Ferodo

6. Aircraft
   SA des Ateliers d'Aviation Louis Breguet

7. Rubber and Plastic Products
   Michelin & Cie

8. Non-Metallic Mineral Products
   Saint Gobain

9. Coal and Petroleum Products
   Cie Française de Raffinage

Switzerland

1. Chemicals & Pharmaceuticals
   Society of the Chemical Industry in Basle (Ciba)
   Durand & Huguenin
   Sandoz
   J. R. Geigy
   Hoffman-la-Roche

2. Mechanical Engineering
   Ste. Sulzer Freres
   Landis & Gyr
   Adolphe Saurer

3. Electrical Equipment
   Brown Boveri & Cie

Summary

PAUL ERKER

Economic history is discovering anew the importance of innovation, science and technology. The changing pattern of innovation and the events and processes which link science, technology and innovation both on the macroeconomic and the microeconomic level are, therefore, the theme of this book. Innovation is seen not only as a new product or process, but as a set of activities, itself a process, which transforms ideas, concepts or scientific and technological potential into commercially available and useful products, processes or techniques. Process definitions like this one are in some ways the most useful if one is considering innovation from a managerial or decision-making viewpoint, because they can focus attention on the critical problem areas on which the outcome depends. It is a definition, in addition, on which historians and economists can agree.

Economic historians like to characterize the interwar period in Europe as a period of relative stagnation. Nevertheless, it was also a period in which numerous innovations occured: radio and television (in its infancy) as mass consumption goods, several important synthetic products in chemistry and pharmacy, radar and the helicopter – to mention only a few well-known examples. One of the basic questions is therefore: how did the interwar period fare as far as innovations are concerned in comparison with the pre- and postwar periods? How were innovations being generated within firms, industries and nations during this period of both strong growth and severe crisis?

The contributors to this volume – historically minded economists and economically minded historians – are not only drawn from a number of different countries, but also from a variety of academic disciplines, with their own methodological views and approaches to the theme. This volume is not, therefore, intended as a comprehensive treatment of the vast subject, and it reflects the particular concerns of each author. None the less, read together, these essays about the emergence of research capabilities, the innovation strategies of firms, the progress of new industries, and technological transfer emphasize several common elements of a complex picture of the interwar period. It was *the* time of expansion of innovations as (large) technical systems (telephone, radio, electricity, traffic), and that means the development of innovation patterns. In addition it was *the* time when inventing organizations emerged and expanded as (small) scientific systems, and in which R&D became the critical factor in business strategy. Economic and political policies changed from affecting to steering the march of technology.

This collection of papers represents not a culmination but an exiting and challenging introduction to the state of the art in technological, business and economic history. Despite the methodological, conceptual and thematic range which they encompass, these essays offer a decidedly new perspective a) on the history of the interwar European economy, b) on the history of the expansion and change of big business, and c) on the history of regional variations of technology.

First, attention is drawn to the fact that in the interwar years, the European economy was not only experiencing a period of instability and crisis. It was also a time notable for technical and scientific advances, the emergence of new products and new industries, and for strong expansion in some countries, e. g. in Sweden, for at least part of the period. From the perspective of a history of innovations the interwar period was characterized on the one hand by the application of known technologies and production processes spread over a wide area. Most of the basic inventions had been completed before the First World War, and were intensively exploited in the 1920s, during which technological change appeared in the old as well as in the new industries. But on the other hand, the main emphasis changed to the new products of durable consumer goods, giving us an impressive number of technological innovations during the 1930s. In the midst of the Depression important innovations occured, some of which enjoyed rapid growth. The Second World War produced a backlog of inventions which took off during the postwar years. During the interwar period the seeds of the boom of the 1950s and 60s were planted and nourished. Thus, with its signs of growth and instability, the interwar period was a kind of experimental laboratory, showing certain highly distinctive features and contradictions. The papers are a plea to end the separation of the interwar period from the rest of the 20th century and to reassess those years by taking the long-term view. Although the period from 1919 to 1939 was profoundly influenced by the First World War, it also inherited and handed down the major trends of the prewar years. In this respect the interwar period created reserves of productivity and a potential for innovations which remained unexhausted and could be fully mobilized only in the years after the Second World War.

Second, whereas we have a macro-economy of disequilibrum and an innovation theory related to technological interface and economic transition which might serve as a touchstone for a rereading of the history of those years, we have, for the history of business enterprises, something like a micro-economy of uncertainty and a theory of innovative capability that might provide a guideline for an understanding of the interwar period. Fundamental changes in the organization of firms occured. Most big firms only then opened research laboratories and started systematic research into new fields. Those years can be labeled as the era of R&D, when the invisible hand of university-based science had changed to the visible hand of industrial R&D. The innovative capabilities developed in R&D laboratories also provided the firms with new strategic features and functions, be it a strategy of expansion, diversification, modernization or cooperation and competition. Innovation was seen as a natural outcome of investment in research and development, and attention was directed at the importance of the organizational style and structure needed for innovation. But the Depression seemed to be a turning point in the managerial belief in R&D as the key to prosperity. The papers show that firms adopted many different approaches to new product development in general, and innovation in particular. They suggest an evolution of the focus about innovation from the discussion of a phenomenon rooted in R&D towards consideration of it as an activity which spans many functional activities, has many organizational and attitudinal barriers to success, demands considerable management skill and has high risks attached. It must be seen much more as a social process, with its inherent interface problems. In addition they plea for a shift towards inclusion of the innovative role of small and middle-sized firms who found themselves niches for new developments, e. g. as suppliers of the big enterprises in the automobile and aircraft industries.

Third, technical innovations and technical change in general can no longer be reduced solely to their contribution to economic development and economic growth. This volume tries to help close the gap between the work done by economic historians and economists on the sources and the development of economic growth on the one hand, and the work of historians of technology on the other which, in approach and outcome, is seldom linked to the former. The papers have shown that one should put greater emphasis on the "costs" of technical change, not so much in the pecuniary sense, but with respect to its effect on the environment including the function of integrating economic areas and of reducing or accentuating the patterns of innovation on national and supranational level. Though not explicitly studied but nevertheless thematisized is the need to give particular emphasis to innovations from small firms or countries and the diffusion from the "centers of excellence" into the "periphery". Some new techniques, such as wireless radio transmission, promoted the dwindling of invention and innovation processes through the cartel agreements. However, the radio as a consumer product reduced the level of economic and managerial control and opened opportunities for other firms, including from smaller countries, as the example of the Dutch electrotechnical firm Philips shows. Indeed, if we look at patent patterns in identifying changing rates of invention and innovation, it seems that there has been a specific European pattern of innovation taking the form of a slow but steady technological development. But the R&D potential within US firms had doubled in that period, whereas within German firms it tripled. The decline in the rate of growth of technological activity in the 1930s characterized Germany and France as well as the USA, but did not extend to Britain. In a way the dicrepancy in the pattern of technological activities between the European countries had become more pronounced by the interwar years.

The papers collected in this volume thus challenge accepted interpretations, suggest some alternatives and at the same time offer a context in which new questions can be posed and answers sought. The interwar years are again open to historical research with a fresh look at the things.

Index

Bei Fragen zur Produktsicherheit wenden Sie sich bitte an:
If you have any questions regarding product safety,
please contact:

Walter de Gruyter GmbH
Genthiner Straße 13
10785 Berlin
productsafety@degruyterbrill.com